OLD MAN IN A BASEBALL CAP

a memoir
of world war II

FRED ROCHLIN

Perennial

An Imprint of HarperCollinsPublishers

TO HARRIET

First Perennial edition published 2000.

Designed by Celia Fuller and Laura Lindgren

The Library of Congress has catalogued the hardcover edition as follows:
Rochlin, Fred, 1923–
    Old man in a baseball cap : a memoir of World War II / Fred Rochlin.
        p. cm.
    ISBN 0-06-019426-X
    1. Rochlin, Fred, 1923–   . 2. World War, 1939–1945—Aerial operations, American. 3. World War, 1939–1945—Personal narratives, American. 4. United States. Air Force Biography. 5. Flight crews—United States Biography. I. Title.
    D790.R645   1999
    940.54'4973—dc21   99-21421

ISBN 0-06-093227-9 (pbk.)

00 01 02 03 04 ❖/RRD 10 9 8 7 6 5 4 3 2 1

"This simple memoir is not your typical story of World War II. Rochlin has written a collection of brief remembrances that he performs on stage as a one-man show, a monologue of life and death as a navigator on a B-24 bomber in Italy from 1943 to 1945. . . . His observations of people and events are keen, ribald, and very funny." —*Library Journal*

"A shining example of how the oral tradition can translate into text. A very good read. I'm proud to be part of Fred's literary achievement."

—Spalding Gray

"Fred Rochlin's unassumingly tender reminiscences of his days spent dropping bombs in Europe manage to invest the crude mechanics of war with unforgettable images of the unquenchable beauty of humanity. I was enthralled, touched, and very deeply moved."

—Simon Winchester, author of *The Professor and the Madman*

"A terrific book, full of sex and violence and honor and kindness. Somehow, Rochlin has managed to keep the kid alive inside him, so the story of World War II reads as if it's being told to us—as it happens—by this young navigator right from the nose of a B-24."

—Susan Isaacs, author of *Red, White and Blue*

"It's a real-life *Catch-22*. I couldn't put it down."

—Michael Korda, author of *Another Life*

"*Old Man in a Baseball Cap* remains a recollection that is about as honest as can be of a young man and a time—and oh what a time it was—which lives now only in a wavy light in the waters of memory."

—*Los Angeles Times*

"In [the book], Rochlin morphs from old man to man-child and back, unseating himself in time—mentally, at least—like *Slaughterhouse Five*'s Billy Pilgrim, yet dodging the staged lunacy of *Catch-22*."

—*Los Angeles Magazine*

"Using a series of vignettes, it's a common person's way of relating the cataclysm of the Second World War and how it affected one nineteen-year-old from Nogales, Arizona, who joined the Army Air Corps, went to Europe, and witnessed things no nineteen-year-old should—and therein lies its power." —*Denver Post*

"His stories of petty officers, unprepared enlisted men, bombing raids gone wrong, and tiny acts of heroism are fascinating. . . . Rochlin's simple prose makes us comfortable, as if we were sitting at a table with an old friend or relative, someone who had lived a remarkable life and now was ready, after a glass of wine or two, to share a bit of it." —*Booklist*

"Rochlin survived fifty missions and made it home, but most of his friends did not. His poignant acknowledgment of their sacrifice and his wonder at his own survival provide a stark glimpse into men at war."

—*Cahners Library Journal*

"This book version of Rochlin's critically acclaimed one-man show of the same name offers a look at World War II that is by turns horrifying, sobering, and hilarious." —*Publishers Weekly*

Fred Rochlin, 1944

CONTENTS

ACKNOWLEDGMENTS

So MANY PEOPLE HAVE BEEN generous, helpful, encouraging, and gracious to me in regard to this work and my theatrical performances.

My profoundest respect and thanks to Laurie Lathem for her direction, patience, sensitivity, and strength.

Thanks to Larry Ashmead, Allison McCabe, Michael Dixon, Valerie Smith, Bruce Weber, Karen D'Souza, Jennifer Poyen, Marilyn Johnson, Mary Jane Ufland, Harry Ufland, Allison Gregory, Steven Dietz, Peter Matson, Jens Hansen, Paul Stager, Larry Silverton, Spalding Gray, Carmen Ulloa.

Thanks to the great people at MacDowell Colony, Highways Performance Space, Actors Theatre of

Louisville, B Street Theatre, Emelin Theatre, Peterborough Playhouse, La Jolla Playhouse, Playwrights Kitchen Ensemble.

My deepest gratitude is due my wife, Harriet, and our offspring, Judith, Davida, Margy, and Michael, for their unstinting support.

THE STORIES WE TELL

THERE ARE A LOT OF OLD GUYS around out there in their seventies and eighties. I'm one of them. Some of us went through a lot of stuff that we have been unable to talk about. We are in the last decade of our lives and sometimes we worry about our approaching oblivion. We are part of that generation that was raised and trained to repress their feelings or to deny and try to forget so many things that occurred.

Everybody has a story. I believe everyone's story is important; should be told, retold, written and recorded. If there are, say, five billion people on this planet, then yes, I think there should be five billion stories told.

We all go through a struggle, the path of our lives. I call it "La gran lucha." The great contest. The ultimate wrestling match.

The stories in this book were written to be performed as monologues in a theater. As such, a purpose was to hold the audience's attention, to entertain and to illuminate. Hopefully, the audience would be moved into thinking about their own lives, their own struggle.

My country, the United States, was involved in this epic fight in Europe called World War II from December 1941 until May 1945. In these stories, I try to show the small personal feelings of a young man tossed against all the other stuff that was going on.

When I meet other old guys who had experiences similar to mine, I'm dazzled by what they say. We are often so inaccurate. We remember what we want to remember. We deny what we choose to deny. We have become a generation of storytellers fixed on that period of time when many of us were so young, experienced so much, were so scared and traumatized.

My memory of those times is a blur. I don't remember places, dates, or even what happened with any kind of exactness. The older I get, the more I remember things that never happened.

The stories that are written here come out of my experiences and are in my own voice. I do want the reader to understand what I'm trying to express. Look, from my perspective, we just get one shot at life. Don't piss it away. That's all I'm trying to say. Tell your story. Tell your story.

MILK RUN TO GENOA

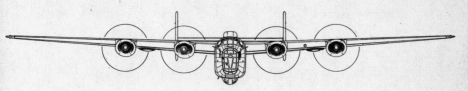

IN DECEMBER 1942, I FINISHED my cadet training, was commissioned a second lieutenant, and was ordered to Mather Field near Sacramento, California, to meet the crew I was to be assigned to.

We were to take an eighteen-week training course in practice bombing, formation flying, aerial gunnery, and navigation, and do all the stuff that you have to learn to get ready for combat.

We started training flying the very next day, and the first week, we just flew all over the United States, dropping bombs on bombing ranges, practicing formation flying, aerial gunnery, weird navigation exercises; we did very well, so well that, by the end of the

3

first ten days, we were the hottest crew on the base and had the best record in everything.

Shorty Haden, the bombardier, was my roommate at the BOQ, the Bachelor Officers' Quarters, and we got to be really good friends. We were about the same age, and we'd just talk and talk and talk.

When we weren't flying, all the guys would get off the base, go to Sacramento or San Francisco to try to pick up girls or at least get drunk. Kind of relax, try and have a good time, you know, what guys are supposed to do.

Well, not Shorty, he'd go to town and spend quite a bit of time in church. He was very religious and then he'd go and haunt these little antique shops.

The first Saturday we had off, we all went to the Wagon Wheel Hotel in town. Shorty said he'd meet us back at the base later.

I got in about midnight, and Shorty was there still awake and waiting for me and he was just gushing, alive with excitement, and he said, "Oh, you'll never guess what I found today."

I said, "What?"

And he said, "Look here," and he opened the lid to his foot locker and there were his clothes and three little cups and saucers.

So I said, "Look at what?"

He blurted, "At these, at these," and he took out those little cups and saucers and said, "Aren't they just precious? Can you believe this one is a genuine Worcester, late eighteenth century? Can you believe how lucky I am? Aren't you jealous? And this is a Miessen, and this is turn-of-the-century Wedgwood, not really rare but still very nice."

I tried to be enthusiastic, I didn't want to be a wet blanket.

I said, "Swell. Where'd you find them?"

He said, "Sacramento is just a gold mine, no one here knows what they're worth."

"Yeah, what they cost."

"Only eighty-five dollars."

I said, "Jesus, that's half a month's pay."

He said, "Yes, but they're worth five times that, these are museum pieces. Oh, my mother will be thrilled when she gets these. And I have to borrow fifty dollars from you till payday."

Well, I thought, This guy is a nut case, but what could I do, so I said, "Sure."

The next night, he took my fifty dollars, went to the Officers' Club, got into a poker game and won one hundred twenty-five bucks. Then he paid me back the dough he owed me, got in another poker game and won two hundred dollars and change.

Then he went to Sacramento and brought back a whole mess load of those damn little cups.

So I said, "Shorty, what are you going to do with all those damn little cups and where'd you learn how to play poker so well?"

Shorty said that he was an only child and his mother was a widow and she loved these little cups, had a collection of them, and she and Shorty studied about porcelain almost every evening after school when he was home. He knew all about Spode, and Rockingham and Royal Danish, and St. Ives, and Rouen and Lowestoft, and every other damn cup.

And he said, "And everybody in Jefferson City plays poker and it's not too hard to win if you understand how, just be patient, don't be afraid to fold, play the odds, only bet on the good hands, and the most important thing, don't expect God to intervene to make you win."

He thought the guys at the Officers' Club were nice guys, but just didn't know anything about playing poker.

So I figured Shorty wasn't so nuts after all, he just had the hots for little cups, okay? Look, it takes all kinds of guys.

After three weeks Major Ferguson, the chief training officer, called in the whole crew. We were worried, didn't know if we were in some kind of trouble.

Major Ferguson said, "Your crew has been here three weeks and you've got the best performance record of any crew on the base. Do you men feel you're ready for combat? Do you think you can fly fifty combat missions?"

"Yes sir." We were getting excited.

"Well, you've only been here three weeks, but you're doing better than crews that have finished the whole eighteen-week course. They're desperate for replacements in the 15th Air Force in Italy. Can you handle that?"

"Yes sir."

Major Ferguson said, "Okay, we'll issue the orders this morning. Get packed up and fly to Wichita this afternoon and pick up a new plane there. Good luck, gentlemen."

That was that.

Outside, we said, "Italy—WOW!" We were excited, we were hoping for England, we didn't want the South Pacific—never thought of Italy.

We flew to Wichita that afternoon, where we picked up a brand-new bomber, a Consolidated B–24. The next day, we flew from Wichita to Miami. The next day, we flew from there to Trinidad. Next day to Natal, Brazil. Next day, from there across the Atlantic Ocean to Dakar, Senegal, in West Africa. Next day, from there across the Sahara, over the

Atlas Mountains into Marrakech, Morocco. And the next day, from there to Bizerte, Tunisia. We were out of the United States six days. At Bizerte we were told to refuel and fly on to Orto Vecchio, Italy. We landed at Orto Vecchio. At the tarmac, a jeep was waiting for us with a big sign that said "Follow Me" and we followed it. We went up to our staging area, got out of the plane, and got on a truck. They took us to the 656th Squadron, we were going to be assigned to it. We met Major Thompson, commander of the 656th Squadron; he took us into his tent, he welcomed us. We were going to replace another crew. He told us what to expect, what to do. He was very nice to us. I appreciated him. He sort of told us the rules, where the officers ate, where the Officers' Club was, where the enlisted men ate, where their club was, where our tent would be.

He said, "If you have to get laid, go to Foggia. Don't go to Orto Vecchio. If you get gonorrhea, we'll put you in jail. If you get syphilis, we'll shoot you. You came here to fly, not to fuck."

He said, "The best thing to do is jack-off. That's what I do."

Major Thompson was straight arrow. He was out of West Point. He told you how it was.

A sergeant took us to our tent.

The sergeant said, "We haven't cleaned this tent

up yet, but we'll come by and get the other guys' stuff."

I said, "What do you mean?"

He said, "This is the crew that was shot down yesterday. This is Tom Harrison's crew."

I thought, "Oh-oh." We went in there, I felt kind of creepy. Their cots hadn't been made. Their clothes were lying around, socks on the floor. There was Tom Harrison's wife's or girlfriend's picture, she was cute.

The sergeant said, "If there's anything here to drink, drink it. Take whatever you want. We'll come after the other stuff later."

That night, we went to the Officers' Club. Major Thompson was there and said, "I'm making assignments right now. We break you in little by little. For tomorrow we need a navigator, a bombardier, and a nose gunner. We'll take Rochlin, navigator; Haden, bombardier; and Hughes, nose gunner. You'll all fly with Lieutenant Carson."

I looked at Shorty.

Shorty said, "Jesus mercy."

I thought, "Less than a week ago, we were still in the U.S." I was excited—real combat, no more training exercises. At 4 A.M. the next morning, some guy woke us and we went to briefing. We were going to go and bomb Genoa. Genoa, Italy.

Major Thompson said, "It's going to be a milk run, the Italians can't shoot straight. There are some German ships inside the harbor. That's what we're going to go after."

Seven planes, one squadron. The American troops had just landed at Anzio Beach, there hadn't been a breakthrough yet. The German Marshal Kesselring had all his troops against the Americans and had his supply ships in the Genoa Harbor.

It was still dark about 5 A.M. and after briefing, a jeep took us to the plane, where the gunners and ground crew were waiting for us. They had a little fire going in a fifty-gallon oil drum. They were all standing around warming themselves. The bombs and ammunition were already in the plane.

We stood around, I tried to act very cool like I wasn't excited or nervous. We stored our stuff aboard. Then everybody stood around the nose wheel and peed on it. That was for good luck.

The gunners asked, "Where we going?" and Carson said, "Genoa, should be a milk run."

Hughes, the nose gunner, Shorty, and I crawled into the nose through the bomb bay. In a B–24 in a combat situation, once you're in the nose, you're isolated, separated from the rest of the crew, can't move throughout the plane without a lot of difficulty, and your way of communicating is through the intercom system. There is just barely space for the nose gun turret with its two fifty-caliber machine guns and boxes of ammunition, the bombardier with the Norden bombsight, and the navigator. It's about the size of a large card table. There's no space, you sit on ammunition boxes, we were all just jammed in. If you're at all a bit claustrophobic, you're in trouble.

The plane took off and I gave Carson, the pilot, the course for Genoa. We hit cloud cover just past Ancona point, from then on we were flying in the clouds, couldn't see the ground. You had to navigate by a system called dead reckoning, so I had to be very careful and know what the winds were. This was before radar. When we got to about twenty-two thousand feet, I had realized that I'd screwed up. It was about twenty degrees below zero and I was freezing. It was awful. I had made a mistake and had stowed my leather flying clothes in the wrong part of the plane where I couldn't get to it, I was worried. Worst of all, I thought, my balls were freezing. I was afraid of losing my balls. Did the stuff inside your balls freeze? I didn't know. Didn't know what to do. Shorty was on one side, Hughes in the nose turret, and I was kind of jumping around and I had to navigate so I couldn't put my hands in my crotch to keep things warm. And then I spotted some gloves, an extra pair of flying gloves that somebody had tucked on a wire between a bulk-head and a strut. I got one glove from there and I stuck it in my pants and I put my balls and cock in it, and I was saved. It worked, it worked. I got it so the glove fit right in the crotch of my pants.

About the right time, I said, "Carson we're about five minutes from Genoa." We were still in the clouds and you can't bomb a ship through the clouds. Ships are a tough target. You've got to see them to hit them, so Carson took the squadron around and we circled down. At about fifteen thousand feet we broke out of the clouds and there was Genoa, about two minutes away. Shorty got on the bombsight. With the Norden bombsight, the bombardier is in control of the plane. During the bombing run, he actually does the evasive flying while he's lining up the target. He's in control. The pilot just sits there. When we were at about fif-teen thousand feet, the Germans were expecting us and they started shooting. And we'd see these puffs of black smoke, "womf, womf, womf, womf" all over, and the concussion made the plane shake and rock and pitch. And I looked at it and those fucken things got closer and closer and all of a sudden, I realized they had our altitude. We were boxed in. And then "wham." The plane just vibrated all over and Carson screamed, "Feather the number two engine. Feather the number two engine."

What happens is, when the engine is hit, you have to feather the prop so that it's going into the wind, otherwise it's a terrible drag and you go into a spin and you've got to do it quickly before your hydraulic system goes out.

Carson screamed in the intercom, "Shorty, can you still drop the bombs? Hurry up, we gotta get the hell out of here." Shorty was on the bombsight and he was still flying the plane and he finally said, "Bombs away." He had dropped the bombs. Carson put the plane into a left bank and then "WHAAAAM!" And my head was snapped back. I was knocked flat against the back bulkhead and I was unconscious. I didn't know how long I was unconscious, but I got up and there was all this smoke and stuff all over. And where the nose gunner had been was this big god-damned hole. Hughes was gone. There was just this big god-damned hole there like three feet from me. I started going back against the spar, I was afraid of falling out of that hole. There was hydraulic oil all over, fifty-caliber shells, and I kept slipping. I couldn't, I couldn't get a grip. I looked over at Shorty and there was Shorty's neck, his head had been blown off, and all that was left was the back part of his skull, and his blood and skin and bone and stuff were all over me. I got on the intercom and I screamed, "Carson."

Carson said, "Shorty, what the hell happened there?" I said, "This is, this is, the navigator. Shorty's dead."

And Carson said, "Oh shit, we've only got the three engines now and we've lost a lot of fuel and I don't know where the hell we are. We've gotta find our way back. Where the fuck are we?"

I looked at my watch and we'd been away from the target for about thirty minutes and I said, "How fast have we been going, Carson?"

He gave me a speed and I said, "What direction?"

And he said, "South."

I started to plot the map to get a fix on where we were and Carson said, "Don't fuck up. Don't fuck up." And I was thinking as fast as I could and I gave him a course. About thirty minutes later, we broke out of the clouds and I could see the coast of Calabria and I knew we'd be all right.

I started praying, but didn't really know how, and I said, "Mama, I don't know what I'm going to do. This is a milk run. Mama, this was the first one. I don't want to flunk out. Mama, I don't think I can do this fifty times. Mama, I'm just not tough enough. Oh Mama, I'm in big trouble—and look, there's poor Shorty. What about his ma? What about those cups? Oh Mama, I don't know what to do."

ROCKETS

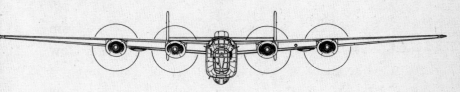

My name is Fred Rochlin.

I was born and raised near Nogales, Arizona.

My parents had emigrated from Russia.

I had two brothers and two sisters.

I was the youngest.

We lived in the country.

We had chickens and turkeys and a black and white Holstein cow named Bossy.

Nogales had about five thousand people in it. It had a school and a library and a city hall and a county courthouse. It was a ranching and mining and railroad center and a border town.

I liked Nogales.

I thought it was a nice town.

I had a summer job working in the stockyards.

In high school, I was sort of a flash.

At graduation, I got to join the National Honor Society.

I went to the University of Arizona and I majored in civil engineering because that's what my two brothers had done.

I thought that was the right thing to do.

When I got there, I found that I couldn't pass anything. I couldn't pass geometric equations. I couldn't pass algebra, couldn't pass calculus, chemistry, surveying, physics, differential equations. I couldn't pass a damn thing. I was flunking out and that would be a big scandal in my family. I was getting desperate.

I didn't know what to do.

That December, the Japanese government saw fit to bomb Pearl Harbor.

So, next month, January, two weeks before finals, I got very patriotic and I went down and enlisted in the Army Air Corps.

Now the Army had this special program for college kids called Aviation Cadets. See, everybody wanted to be a fighter pilot and fly a P–38.

So they had this sorting-out process and gave you these aptitude tests. Then it turned out to be kinda simple. It seemed like all the kids that were

WASPs were sent to pilot training. Kids with Scandinavian or German last names to bombardier school. All the Catholic kids, the Irish, Italian, and Polish kids, right to gunnery school. Mexican kids were given a choice between infantry or paratroops.

Then there were the Jewish kids. What to do with them?

Well, hell, there were only ten of us and every single one of us was sent to navigation school.

Our drill sergeant explained, "Look, everybody knows you Hebes are good with numbers, and all a navigator is, is just a fucken flying accountant. You know, a guy with a briefcase."

Eight months later, I had finished my navigator training and was sent to meet my crew. We were gonna fly a B–24, a four-engine heavy bomber.

I met our pilot, he was a nice guy from Montana. We got to be friends. His name was Bill Rickets and that's when they started calling me Rockets. See, Rickets and Rockets. That sorta fit, kinda rhymed, it sounded cute. And that's what we were, cute, just a cute crew.

We were ordered to join the 15th Air Force in Southern Italy. And when we got there, we found our job was to bomb the southern perimeter of Europe and bomb in front of the advancing Russian armies that were coming in from the east.

I'd flown a mission that day and I was tired. These missions took eight, ten, sometimes twelve hours. I went to the Officers' Club. The Officers' Club was a little tent, a pyramidal tent, ten feet by ten feet with a table in the middle. And I was drinking brandy in a twisty blue Italian glass, it looked like it had been made in Mexico, and I was proceeding to get drunk. In came Captain Bill Connor, the flight surgeon, a guy from Boston, a nice guy, he liked me, and he said, "Hey Rockets, you speak Italian and I need a translator. Want to help me deliver a baby?" I didn't speak Italian, I spoke Spanish, but no one knew the difference or cared, I was half drunk anyway, and I said, "Sure, why not?"

I got up and got in his jeep and off we went to Orto Vecchio, this little farming village that was near the air base. Orto Vecchio had about thirty stone houses. They were stone all the way up to the top. There was one main street. The street was shaped in sort of a V-shaped channel and was full of sewage. There was a dead horse at one end that hadn't been moved for a month. The village smelled so bad that when you got there you breathed through your mouth because you couldn't stand the smell.

We pulled up to this one little house and there was this guy standing in front waiting for us. It was Mr. Carafa. I recognized him, he was the guy that brought vegetables to the air base and sold them to us. We stepped into this little house, it was dark, but there were three little kind of oil lanterns. Off to this one side was a fireplace, a stone fireplace, and standing there was this woman, a middle-aged woman, obviously Mrs. Carafa all dressed in black, round shouldered, hawk nose, and fierce-looking eyes. And then, in the middle of this room, was this table, a wooden table, kinda greasy, and from the ceiling it was all garlic. It had been braided and hung up there to dry. Again, the odor was just overwhelming. I thought it was toxic. And off on this side, on a pile of rags was this body, and Connor and I went over and there was this girl, obviously swollen, and her arms were dark and her legs were dark. At first I thought she was one of these kids that had come over from North Africa with the troops.

Connor reached down into his bag and pulled out a sheet, a white sheet, and spread it out on this table, and then he and I went over and lifted up this gal and got her on the table, and then he said, "Rockets, you tell Mrs. Carafa that I have to take off her dress to examine her." So, I said, "Señora, necessita sacar este vestido."

And Mrs. Carafa looked at us and just shook her head.

So, I said, "Mira, el es un medico, Americano, muy bueno." He's an American doctor, he's very good. Then she stared at Connor and finally nodded.

So Connor pulled up this gal's dress but he couldn't get it past her belly, her swollen belly. So he reached down into his bag and pulled out a pair of scissors, those blunt-nosed, surgical scissors, and went "whoosh" and just ripped that rag right apart.

I was shocked. You see, I was very young and although I had had those usual teenage sexual forays in the backseat of a Chevy, I had never really seen a naked woman before. And there was this woman, this fourteen- or fifteen-year-old woman, with this huge, huge belly and these tremendous breasts. And each breast, each breast had these big kind of rosettes that almost covered the entire breast and each, each rosette had this huge kind of nipple on it, looked to me like two upside-down flower pots. And then her skin was like dark blue and red and purple and magenta, all blotchy. And Captain Connor said, "Jesus Christ. I think she's got peritonitis, we've got to save this gal. We've got to do something." He said, "Tell, tell Mrs. Carafa to warm some water. We're going to have to wash her first." Then he put on a pair of rubber gloves and he groped between her legs and he mumbled something about dilation. I didn't know what the hell that was. Then he said, "Rockets, we're going to have to do a cesarean right here, right now. We're going to have to try to save the baby. We may not even save the mother." He handed me the scissors and he said, "We're going to make an incision just about here. Start cutting her pubic hair."

I looked at that pubic hair. She had hair from her belly button all over her belly, all around her hips, down her thighs all the way to her knees. It was a mess of hair. It reminded me of a little black bear I had seen once in northern Arizona. It was silky, kind of oily. I started cutting, I started cutting that hair. Connor reached into his bag and got out these little vials of morphine. They come in these like little toothpaste tubes with a needle on the end. And he started sticking them into her side and squeezing out the morphine. He put three on each side. That was a lot of morphine. And then he lifted her head up, he put some kind of pill in her mouth. He didn't give her any water, just some kind of pill. The girl all this time, at first she looked kind of terrified, big eyes, never said a word, didn't murmur. Then she seemed kind of relaxed and pleased that she was getting all this attention. Connor looked down at me and said, "For Christ sakes, hurry up. Get the mother to help you." And I said, "Señora, ayudéme. Necesitamos cortar todo este pelo. Cerca de la chucha." And then I was embarrassed. I said, "chucha." Chucha is the Mexican word for cunt. You shouldn't say cunt to a woman in front of her daughter. That wasn't nice. And then I worried about the grammar, is it "el pelo" or "la pelo"? I couldn't remember which, but the mother, she understood and she started helping me cut it.

Captain Connor got this warm water, washed her belly, dried it, got a bunch of iodine and washed her belly with the iodine, let that dry. Then he pulled out a scalpel and went "whoosh" right across her stomach between her navel and her vaginal opening and all this water and fluid and blood just gushed out and he took two clamps, pulled the skin back and said, "Rockets, you hold these," and put two more clamps and the mother understood and she held those. He then groped into this woman's belly and it seemed to me, messed around, and then pulled out a baby. I had worked the stockyards. I'd never seen anything like that before.

It was a little boy, pink face, pink body, and the baby started crying. Connor took the baby, tied the umbilical cord, cut the umbilical cord, washed the baby's face with warm water, put some drops in the baby's eyes. He reached down and got a sheet, a clean white one, wrapped the baby in that and gave it to Mrs. Carafa, now the new grand-mother. And she was smiling.

He then reached down into the cavity and pulled something out and said, "Here Rockets, take this," and he gave me this pile of goop and I didn't know what to do with it, so I just dropped it on the floor. That was okay. It was a dirt floor. I wiped my hands. Then he picked up some gauze, went down in there, mopped up a lot of blood and handed me that. I threw that on the floor. Then he sprinkled in some white powder. I didn't know what that was. I learned later it was kind of a new drug we had called sulfa. He had a needle and thread and went down there and he made some stitches, sewed something up, got rid of the blood some more, handed me the gauze, pulled back the skin, poured on some more sulfa, stitched that up, put on a com-press then bandaged up the whole belly, and he straightened up and was smiling. He was done.

Connor, I always thought was sort of a crummy flight surgeon, obviously was very competent. I was proud of him. I thought, "Good for you, Connor, and good for the American Army. We've done the right thing." And we started to leave. The girl was asleep. Mr. Carafa gave us a shot of brandy. We said, "Salud." Connor was covered with blood, from top to bottom, from his shirt to his pants. And every time he took a step, his shoes went "squish, squish." They were full of blood. We started to leave and when we got to the doorway, Mrs. Carafa stopped us and said, "Come chiama il medico?" and I said, "Guglielmo," and she said, "Voglio chiamare questo bambino, Guglielmo." And I said, "Connor, they're naming the baby after you." He just smiled and shrugged. Then she said, "Quale é il nome secondo del medico?" and I said, "Connor, what's your middle name?" He said, "James," and I told her, "Iago." She said, "Voglio chiamare questo bambino, Guglielmo Iago." We started leaving, she got a hold of me and said, "Quale é suo nome?" and I said, "Federico." She said, "Il nome del questo bambino é Guglielmo Iago Federico Carafa." Well, no one had ever named a baby after me before. I was moved, I leaned over and kissed her, kissed her on the cheek.

We got into the jeep and we got back home. Home was a little tent. I on my cot. See, after every mission, they give you these woofer pills and you take them with a shot of whiskey and it helps you go to sleep. It's really a pill called Seeconal or Seconal and the reason why they call it a woofer pill is when you take it with the whiskey you just automatically go "woof, woof." Well, I had the pill and the whiskey. I don't know why, but I just tossed down the whiskey and went sound asleep. It was about two o'clock in the morning.

About three o'clock, some guy came and started to shake me and said, "Lieutenant, wake up, you've gotta go fly," and I said, "Bullshit, I flew yesterday, I don't have to fly today." He said, "Yeah, you do." He says, "Captain Smith, the lead navigator, is sick with the flu and the Colonel, he asked for you." Well, I didn't mess with any colonel, if he wanted me, he had me. I got dressed and went up to briefing, didn't shave, and I knew I was going to suffer that day because that meant you had eight hours flying in the plane with the face mask rubbing against your stubble and it was gonna hurt.

At briefing, I learned we were going to go bomb a place called Hadju-Polgar, I had never heard of the place before. It was in Hungary and the British Intelligence had said that it was a gathering place for all kinds of German tanks, railroad cars, munitions and supplies and we were going to have to go out and bomb it that day. It was near the Russian front. I got my winds and my maps and then learned that the main air force, about a five-hundred-plane air force, was going to sweep up and hit Munich that day and bomb the BMW factory in Munich. Then, on the way up there, we were going to peel off and go bomb Hadju-Polgar.

I was glad we weren't going to Munich. That was a terrible target. The Germans didn't like to get bombed, they would defend the BMW factory with everything they had. Five hundred planes were going to go up there, we were going to lose fifty planes that day, that's five hundred guys we're going to lose. I'm glad we didn't have to go.

I have always hated that BMW. You know, out in the parking lot right now, there are BMWs. I hate that fucken car. I see them on the freeway, I give them the finger sign. Do you know how many American boys died over that BMW factory, I think it stands for the Barbarian Motor Works. Himmler drove a god-damned BMW and we're driving them today in America. It makes me sick.

I do drive a Mercedes, but that's different. See, Mercedes were made in Stuttgart and it was too far for us. And the Eighth Air Force in England, they bombed the Stuttgart plant and so that's their god-damned business. And I owned a Leica camera. It's a hell of a camera made in Wetzler, Germany, but the Canadians bombed Wetzler and that's their stuff. But I just can't stand that fucken BMW.

The air force takes off at about five o'clock in the morning. There's five hundred planes, it takes like two or three hours to rally all these planes. The Americans were clever. We'd pick a spot that's easy to find, like say, Venice. You'd find it easily and you'd rally around it, waiting for all the planes to come from all of southern Italy, and maybe North Africa, from Corsica, from Sardinia to get there. Five hundred airplanes. All this time, the Germans would be watching you. They knew what was going on, they had radar. But they didn't know where you were going to go. They didn't know whether you were going to go to Munich, maybe into Czechoslovakia to bomb the oil refineries, maybe even swing over and bomb oil fields in Rumania. They didn't know what you're gonna do. And they had to keep their defenses in reserve until you committed yourself. Well, that morning, off went the air force, five hundred fucken planes to bomb Munich, and we peeled off twenty-eight planes, one group, to go bomb Hadju-Polgar.

A bombardment group consists of four squadrons of seven planes each, one squadron in front, one squadron in back, one squadron on the left, and one squadron on the right, sort of like a box flying. Flying over Hungary was like flying over Iowa, this flat land, these little checkerboard farms, hundreds of them all over. And after about two hours and forty-five minutes, I spotted Hadju-Polgar, and I looked at it, this little old town. Kinda reminded me of Nogales, only smaller, maybe twenty-five hundred people.

I got on the intercom and said, "Colonel, that's Hadju-Polgar, right ahead of us. We'll be there in five minutes."

The Colonel said, he always talked through his teeth, he couldn't open his damn jaw, he said, "LIEUTENANT JENSEN," that was our bombardier, Johnny Jensen right next to me, "YOU GET THAT TOWN IN YOUR BOMBSIGHT."

Johnny said, "I've got it in my bombsight, Colonel, but there's a problem."

"WHAT'S THE FUCKEN PROBLEM." Johnny said. "There's no railroad there in town."

The Colonel said, "WHAT, ROCKETS, YOU FUCKEN SON OF A BITCH, HAVE YOU SCREWED UP?"

When I flew with the Colonel, I never screwed up. I knew where I was every god-damned minute, I was careful. I knew all about this lunatic. I said, "No, Colonel, that's Hadju-Polgar. I am certain."

"LET ME TALK TO LIEUTENANT GOLDBERG, GOLDSTEIN, OR WHATEVER THE FUCK YOUR NAME IS."

He was talking about Bobby Ginsburg who was the lead navigator for the 657th that was flying over here on our left-hand side.

"WHAT TOWN IS THAT, GOLDBERG?"

"It's Hadju-Polgar."

"LIEUTENANT LEVY, WHAT THE FUCK TOWN IS THAT?"

Dave Levy was the lead navigator of the 658th over here on our right-hand side.

Dave Levy said, "That's Hadju-Polgar, Colonel."

"YOU FUCKEN JEWS, YOU ALL STICK TOGETHER. YOU COVER FOR EACH OTHER. I DON'T TRUST ANY OF YOU KIKES ANY MORE THAN I DO A FUCKEN NIGRAH. ROCKETS, WHAT'S THE NAME OF THAT GOD-DAMNED MORMON NAVIGATOR WE'VE GOT BACK THERE."

The 659th squadron was right behind us. It was the fuck-up squadron. The whole idea was that you always kept the lousy fliers in the back because they were the guys that the Germans picked off first. You didn't want to lose your good guys, you want to lose your lousy guys. The lead pilot of the 659th was a guy named McBride. He was the lead pilot and his navigator was a kid from Arizona. I had gone to school with him, we went through cadet training together, he was my good friend. His name was Billy Carswell. Billy never knew where he was ever. He was the most confused navigator in the air force, but we were good friends. We had kind of like a secret code, he would call over the intercom and say, "Rockets," and I knew who was talking, I'd recognize his voice. I would give him coordinates of where we were, but I'd give them to him backwards because I didn't want anybody to know. Like if we were forty-eight degrees north, I'd say eighty-four degrees south. He knew what I was talking about, then he could find out where he was. The Colonel said, "CARSWELL, WHERE THE FUCK ARE WE?" He said, "That's, that's, that's Hadju-Polgar, sir."

I had a lot of respect for the Colonel. See, I knew what his trouble was. What happened is that he was VMI, Virginia Military Institute, he was a professional soldier and we were all a bunch of guys enlisted for the duration of the war. But, this was his career and we had fucked up. We had fucked up. We were the outfit that bombed the Polish, English, and some American troops in Monte Cassino. We had bombed our own guys. It was a terrible mistake, but we had done it. Air Force headquarters was livid and wanted to have someone's head, and the Colonel went up there and took all the heat. Maybe it ruined his career. He wasn't even on that flight, it was Major Thompson that was the lead pilot. But the Colonel, he did it. He stood up for us, he protected us. Then he had these other characteristics. He wore this goddamned shiny varnished helmet and riding jodhpurs and these riding boots with spurs and carried these two nickel-plated pearl-handled pistols like Wyatt Earp, and carried a riding crop and then he would stomp around the air base pretending he was General Patton.

And then one day, a sergeant inadvertently walked into the Colonel's quarters and saw that he was dancing with his orderly and then the rumor started that maybe the Colonel was queer, and that didn't help his career none either.

And then the Colonel was from the South. He was from Mississippi and his daddy had been a member of the Ku Klux Klan and he was raised to just hate black people.

Now this was a segregated Army. And the Army had gone into Tuskeegee Institute and Howard University and picked the best and the brightest black kids and trained them to be fighter pilots and then they put them in these P–51s. These magnificent planes that could go anywhere and do anything and that formed this fighter squadron and they flew fighter protection. They defended us. They protected us.

Well, the Colonel couldn't stand the idea that we were being defended by a bunch of black pilots.

We would get in the air and the Colonel would say, "WE'RE OVER ENEMY TERRITORY, WE'RE GONNA OBSERVE RADIO SILENCE."

The black pilots would be flying along and pretty soon you'd hear, "Hey, who say dat?"

Then you'd hear another guy, "Hey, who say dat, who say dat?"

Then, pretty soon, "Who say dat? Who say dat? Who say dat?"

And on and on and on. All these "Who say dats" through the air.

The Colonel would have a fit. "THIS IS COLONEL SELBY B. SHOOTS. THE NEXT FUCKEN NIGRAH THAT SAYS SOMETHING, I'M GOING TO HAVE HIS BLACK ASS IN COURT-MARTIAL. DO YOU HEAR ME?"

Quiet.

The planes would drone "druuuuummmm." Vapor trails.

Pretty soon you'd hear, "Who say dat?"

"Who say dat? Who say dat?" Then they'd start laughing, making fun of our Colonel.

And he was peculiar. Well, he had to make up his mind. Hadju-Polgar was obviously just a little old Hungarian farming town. British Intelligence had just made a mistake. It had no military value.

He had to make up his mind, do we bomb them, or do we go back home? Then we'd have to drop the damn bombs in the Adriatic. You can't land with a full bomb load. But, the Colonel always said,

"When in doubt, you follow orders, can't go wrong that way."

We were at twenty-two thousand feet and he took us down to eight thousand feet. At eight thousand feet, with the Norden bombsight, you can't miss. And we were the first squadron that went in and Johnny got on the bombsight and dropped those bombs and we just blew the hell out of everything. Destroyed every building, every street, every light post, everything. There was nothing left to bomb. But there were three squadrons right behind us.

And the next squadron came in and dropped their bombs on top of where we had dropped, and then the next squadron came in and dropped their bombs on top of where we had dropped, and then the last squadron came in and dropped their bombs on top of where we had dropped. It took about fifteen minutes and we turned and headed for home.

Look, every plane carries about ten thousand pounds of bombs; ten thousand times twenty-eight planes, that's like two hundred fifty thousand pounds of bombs. Divide that by about twenty-five hundred people, that's like one hundred pounds of TNT for every man, woman, and child in that little town.

Hadju-Polgar was no more. It was gone, finished, finito, a hole in the ground, a pile of Hungarian dust, an ex-Hungarian town, wiped off the face of this earth.

We landed.

I got debriefed. I went back to the Officers' Club and was having a brandy. I had been up for about forty-eight hours and I was tired.

Captain Connor, the flight surgeon, came in all smiles and said, "Hey Rockets, I just came back from the Carafas and the baby's doing great and the mother is just fine and the Carafas are happy as hell; they even asked about you. They wanted to know where was Padrino Federico. They made us both godfathers."

Then he looked at me and said, "What are you so down in the dumps about?"

I said, "Connor, I'm feeling kinda weird. You know how you went out last night and delivered that baby and I guess saved the mother's life, and I helped. I was sort of proud of that. And then this morning, we went out and I guess we killed about two thousand five hundred to three thousand people. And I helped. Connor, isn't this insanity? What's going on?"

Connor looked at me and said, "How many missions you got, Rockets?"

"Fifteen down, thirty-five to go."

"You guys just make me sick. Here you've been bombing the shit out of Europe for fifteen missions and you just catch on you're killing people. What the hell you've been doing, Rockets? You want to be put in a strait-jacket, booby hatch, sectioned eight out of the Army?"

"No."

"Then you listen to me and do what I tell you to do because I'm right. First, you follow orders and do what you've been trained to do.

"Then just forget it. Forget it. And if you can't forget, then pretend. Pretend. And if you can't pretend, then deny, deny, deny. And that drink you're having, finish it and have another and another.

"You do that, Rockets, and you'll be all right."

Well, Connor was right. I did what he said, I did my best to forget. I pretended. I denied and I sure as hell did my share of drinking and I was all right. I thought I was all right. Yeah! I was all right.

LIEUTENANT
DAN CALLAHAN

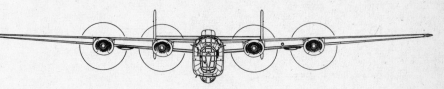

AFTER WE WERE COMMISSIONED, we could live off the base if we wanted, but housing was hard to find in San Antonio. Just about all of us lived at the BOQ, the Bachelor Officers' Quarters.

Not Danny. I don't know how it happened. He met two girls that worked at the A&P in downtown San Antonio and then moved in with them.

He never talked about it, but one of the girls knew another girl, and she told another and pretty soon the word got out that Danny was banging them both. The son of a bitch, he had a harem.

Captain Bobby Brock asked him, "Was it true?" Danny said, "It's not anyone's business but my own, and why should anyone care who I live with?"

Brock was Danny's pal and he quietly listened, and it came out that Danny was in love with both of the girls and wanted to marry them both but that was against the law. Danny and both the girls thought it was a silly law. If all three of them thought it was okay, why should the law give a damn.

They debated, then they thought they shouldn't get married; Danny said maybe that was for the best. "Look, you gotta face facts," he said, "there's a war on and a lotta guys just get killed in a war, right? That's the way it is."

In a quiet way, Danny became the object of envy of most of the guys in the BOQ. I asked him, "Dan, how do you do it? How do you manage to get hooked up so fast?"

Dan said, "You know, I wonder about that myself. Sometimes I think that God just wants to be nice to me. It's like magic. These wonderful women just appear.

"I know I just love women. I love listening to them, what they're doing, what they're thinking about, their problems, their ambitions. I don't lie to them. I don't cheat on them, and I don't make too many promises I can't keep. And I love touching them, smelling them, tasting them. They're all so different. And I'm never the first to make a pass. I love having sex with them, but if they don't want to have sex that's all right.

"I don't know, that's about what I do."

We were all ordered overseas to join the 15th Air Force in southern Italy. Everything was very new to us. We were like a bunch of tourists. The engineers had come in, mowed down this orchard, put in a landing strip. Then we landed, put up tents and moved in.

We were near this farming village called Orto Vecchio, dirt poor, maybe thirty stone huts occupied by peasants who worked the land. This land had been run over by the Italian Army, then the Germans, then the British and Polish troops, and now the Americans. So—they were used to soldiers.

Danny was driving this jeep, and this young kid, maybe a thirteen-year-old girl, kneeled down in the middle of this dirt road, hands clasped under her chin, as if to pray. She was crying. Danny told me he stopped the jeep and how this dirty skinny kid wearing a threadbare black dress was crying and talking and gesturing that she was hungry. Danny gave her a candy bar. She took him by the hand, and led him to this miserable broken-down abandoned stone hut, roof all fallen in, no windows, no door, and there inside was this pathetic young mother with three other young children, two little girls and a babe in arms; dirt floor, a pile of dirty straw to sleep on, a small fire burning on the floor, and a pot boiling. Danny said he thought they were boiling grass.

Danny drove back to the base and then returned to this hut, all in about thirty minutes, but this time with C-rations, canned corned beef, canned sausage, powdered milk, chocolate bars, soap.

He just left it there, and went away, didn't say anything. In five days he made five trips like that.

On the sixth day, he asked me to go with him. He wanted me as a translator. I spoke Spanish, that was close enough. He said he couldn't understand what they wanted. We got there and the hovel was worse than what Danny had described. The face of wretched poverty and despair.

The children were happy to see Danny. They gathered around him as he handed over a box of food.

The mother's name was Mrs. Torcelli. Her husband had been killed. She wasn't from Orto Vecchio but from some other village and was trying to get to Foggia, about fifteen miles distant, where she had relatives that might help her. They were walking, had no money, nothing, and were almost starving when Danny stopped his jeep. She said Danny had saved her life, but then she squinted her eyes and looked down on the ground and said she knew what Danny was up to and they had talked about it and it was okay, Danny could have her or the oldest daughter.

Danny's face dropped and he had me tell her, "Your daughter is just a child and you should be ashamed of yourself, but I won't scold you because these are difficult times," and then he gave her some money and said, "Buy some clothes, and send the children to the nearby church and try to find a place to live. I'll pay for everything until you get on your feet again, and find your relatives. I'll talk to the village priest. I will make arrangements."

Danny gently said, "Don't give yourself away for some canned food. Soon the war will be over. Men will come back. Find yourself a good one. Americans will be gone. Don't do anything you will regret later on."

This village, Orto Vecchio, was a very small place and word of what happened spread like wildfire and Lieutenant Dan Callahan became like a saint.

And about this time, Dan was becoming more and more religious. He'd go to Mass in the village church. The church was a very small affair. The local peasants, the parishioners, were very small people, and Danny being the only person in uniform and besides being the only American, was about six feet, two inches tall. He stood out like a sore thumb. He always sat in the back pew and was very quiet. After services, he would talk to the priest, Father Pietro, who spoke a bit of English.

On the base, the tent where Danny lived was next to my tent and both of our tents backed up to this dusty dirt road.

One day, it was a little past lunch, Danny called and said, "Rockets, come on over here. I need a translator. What's this gal want."

I went into Dan's tent and here was this nervous young woman, I'd judge in her early thirties, neatly dressed, in cheap clothing, too much makeup on, a funny hat on her head, high heels; I quickly noticed her very hairy legs; she looked grotesque, I thought.

Dan said, "Ask her what she wants."

I asked her. She started to talk calmly and matter-of-factly.

"I am from Foggia. I am not a prostitute. But I am here on a mission. The lieutenant here is a young man. A young man has to get fucked or he will go crazy. I am a fucker. I want him to come with me to my place."

I said, "Look, I am a young man too. What about me?"

She said, "You are only a translator. That man is a saint, a savior."

Then she went on. She was a cousin of Mrs. Torcelli; she knew what Danny had done and all the villagers, including Father Pietro, decided that Lieutenant Dan Callahan should be rewarded. But how? The Americans had everything material. What to do? They decided that Lieutenant Dan Callahan needed a woman. But not a prostitute. The lieutenant was too pure for that. They decided on Mrs. Torcelli's cousin because she was a widow, young, lively, and took lovers from time to time, secretly. In the confessional booth she had confessed to Father Pietro just how lonely and horny she was. So Father Pietro figured "what the hell" and gave it his stamp of approval.

Dan told me to tell her thanks, but no thanks, he couldn't accept. And please tell the villagers that getting fixed up wasn't necessary, he could find his own girl.

The lady lowered her eyes, her name was Angelina Torcelli; she said okay, she understood and please accept her apology, she knew she wasn't young enough or pretty enough for the American savior. And the villagers wouldn't give up, they would still keep looking and find someone more suitable. However, just in case he should change his mind, she would leave a note with her address. She would wait until ten o'clock that evening before she locked her door.

On her way out, she thanked me politely for translating; I gave her a plaintive look, she said, "Okay Junior, I'll try to find a fucker for you too."

That evening, Dan checked out a jeep from the motor pool and was gone all night.

He asked me to keep my mouth shut. He quietly went about setting up another place of residence. He got another toothbrush, razor, change of clothes, personal stuff, and he moved in with Angelina. He spent as much time there as he could, and he was learning lots of Italian.

It was early April. The weather was getting better. When the weather held out, we'd fly every other day. One day on, one day off. The Russians were advancing from the East and we were bombing German supply lines, oil refineries, railroad yards, stuff like that. The Germans didn't like getting bombed, and we were suffering tremendous losses, somewhere between ten and twenty percent. Friends, guys you trained with, acquaintances, guys you didn't even know that well, would just disappear; shot down, their tents empty, then new replacement crews would show up. We were all getting tired. Headquarters called it maximum effort. We called it maximum exhaustion. With all this, everybody was getting nervous, jumpy as hell and mostly superstitious. We all started doing nutty things. I wore my wristwatch on my right wrist on all even days and on my left wrist on all odd days. On every Thursday, I changed my socks twice a day. I tied my shoelaces in a particular way for good luck. Before takeoff, before we'd get in the plane, we all stood around and peed on the nose wheel. That was a given, it was mandatory.

All chaplains of all the religions were busy, busy, busy. No one was taking any chances about tripping up God, or getting God pissed off at us.

Danny was quietly crossing himself almost every chance he got and was mumbling more and more to himself and sometimes they seemed to sound like Hail Marys.

Lent was coming up and Danny had to give up something for Lent; he told me he decided to give up drinking Scotch whiskey for Lent. I laughed and said big deal, some sacrifice, there wasn't any Scotch on the base anyway. Danny smiled and said, "Yep, I know." Then he stopped smiling and looked worried.

Later, he told me that he went to Father Pietro and told him what he'd given up and Father Pietro scolded him and said, "No fair, you'd be cheating God. What do you regularly drink on the base?" And Dan said, "Bourbon." Father Pietro said, "Then you must give up bourbon."

Danny said, "But I love bourbon. You don't understand. I've got to have it." Father Pietro said, "Look, give up bourbon, you'll survive. Drink grappa or brandy or wine." Dan said, "Okay," and that's how that problem was resolved.

By the middle of May, we had suffered so many losses that they'd make up new crews for almost every mission. You really never knew who you would be flying with next, but they would try to pick guys who knew each other.

On May 15, the target was the synthetic oil refinery in Germany, called Blechammer. I had been there before and I hated the fucken place. It was really too far away from us. We had to carry too much fuel. You had to deal with enemy fighters and enemy flak. Our own fighter protection couldn't go that far. On our plane, they couldn't put together a full crew. Bobby Brock was the pilot, Danny Callahan, co-pilot. We had a ball turret gunner, a tail gunner, only one waist gunner, no top turret gunner, and no radio operator. I was in the nose all by myself. I was to be navigator, bombardier, and nose turret gunner that day. If we got picked on by enemy fighters, we'd be in tough shape.

We got briefed, took off, got into formation, crossed the Alps, got to Blechammer, spotted the target, and then the shooting started. The Germans must have brought in more anti-aircraft bat-teries that we didn't know about. There seemed to be flak everywhere, those puffs of black smoke. Just before our drop, an explosion rocked our plane, one of our engines had gotten hit. Bobby, the pilot, feathered it, we got on the bombing run, I dropped the bombs, we turned for home. We couldn't keep up with the rest of the formation. Bobby called on the intercom and asked me to plot a course for Switzerland, just in case, and check the fuel. "Can we make it home?" I checked and said, "Yes."

Just then this German Focke-Wulf fighter came tearing past us from up above, shot us up, but didn't turn to finish us off. I didn't know why, maybe he was low on fuel also. The plane started yawing back and forth and Danny was screaming in the inter-com, "Rockets, get up here. Rockets, get up here right away. Bobby's been hit." We were losing alti-tude fast and the plane was bouncing almost out of control.

Look, to get from one part of a B–24 to another part at twenty-two thousand feet is no easy matter. You've got to push ammunition boxes out of the way and they weigh a lot, and you have to discon-nect yourself from your oxygen line and put on a portable tank, and you gotta lug your parachute with you, and all through this narrow catwalk. I got up to the pilot's compartment.

God, what an awful mess. Bobby had caught it right in the chest and parts of him were splattered all over.

Danny's eyes were huge and looked terrified and he was crossing himself.

I got Bobby out of the pilot's chair and laid him on the deck. There was nothing that could be done for him. He was dead. I got in the pilot's chair and hooked into the intercom. Danny was talking and in a firm, calm voice told the gunners in the back to check in. "Was everything okay?" The gunners wanted to know what happened. Danny, in this surprisingly sober voice, said that everything was all right so far. Bobby had been hit, we can't maintain altitude, and keep your parachutes handy just in case.

Danny said, "Rockets, you stay up here with me and be co-pilot. I'll tell you what to do."

I said, "I gotta go back to the nose and get my maps but I'll be right back."

I made that damn trip again, had to crawl over Bobby's body on all fours. Back in the pilot's seat, Danny said, "Give me a heading, Rockets." We all knew the Alps were coming up. It was a clear day. You could see them on the horizon. I checked our position, checked our fuel, checked our altitude, and gave Danny a heading for the Brenner Pass. After a bit, there were mountain peaks on both sides of us. The gunners in back were calling in, "Are we gonna make it? Are we gonna make it?"

Danny, in the most self-assured voice, said, "Now listen. This is Lieutenant Daniel J. Callahan talking and I want to assure and guarantee this crew that we will absolutely make it back, with hardly any more problems. Just be alert."

I thought, Oh, Danny. You're such a bullshitter.

The plane had three engines but one was smoking, losing oil, we were low on fuel, we were losing altitude, and the plane was hard to control, it was rocking and yawing.

I checked my parachute and harness and looked at the escape hatch.

We just skimmed over the snow-capped Alps and there in front of us was the boot of Italy and the Adriatic Sea to our left.

Danny, in this ice-cold, firm voice, said, "Okay, we're over the worst of it but we have to save fuel. Start throwing out everything that isn't nailed down: machine guns, ammunition, flak suits, everything. The chances are good we won't need them."

We knew the Allied Armies had come up the boot of Italy as far as Ancona point. If we could just get past the front lines, we'd be okay.

Danny's guarantee, his coolness, had done its job. Everyone was calm, doing what was necessary. Danny's attitude was infectious, we all became braver.

We got on the radio and started screaming, "Mayday! Mayday!" and called out the coordinates of where we were and asked for the closest landing field.

After a bit, a response came in. They gave us a fix. We were closer than we thought. We spotted the field. It was a short fighter strip. We dropped our landing gear. We landed. We had made it.

It was a Royal Canadian Air Force fighter base. The ambulance came and took away poor Bobby. The Canadians took us to a tent and poured us all a drink.

Up to now, good old Danny was like a rock. He tossed down a whiskey, and started shaking almost uncontrollably, almost like he had a bad case of Parkinson's disease or something. He didn't want anyone to see him like that. He took another drink, and then another and another. Then he settled down.

We called our base on a radio phone. The next day, a plane showed up with our squadron C.O. We

OFFICER
OF THE DAY

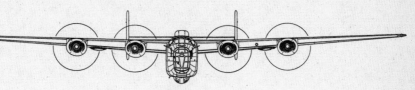

WE LIVED OUT IN THE COUNTRY. The night I was born, it was raining like hell. And my father realized he'd better find someone to help my mother.

He drove to the town and there, huddled under this awning, wrapped in a serape, was Maria, a young Mexican Indian woman, maybe sixteen or seventeen years old.

He brought her home.

She said she was lucky to be in a house with a newborn baby.

She raised me. We became inseparable. She loved me and I loved her.

Maria taught me a lot about life. I remember walking

with her in the desert. She would point and say, "This is rabbit shit. This is coyote shit. This is deer shit."

Then she'd say, "See, shit is part of life. Where there is no shit, there is no life. Just death and emptiness. So don't be afraid of shit. Learn to identify it and then accept it for what it is."

When I was about nine years old, the immigration officials came and took her away.

It was like a death in the family.

I cried for a month.

Ten years later, I was nineteen and I was in the Army Air Corps, and I thought about Maria and her advice and all the different kinds of shit I was encountering in the Army. Could I identify it and then accept it for what it was. Could I?

To me, the Army seemed to be composed of only two kinds of guys. First, the men, the men who understood responsibility, who accepted it willingly; saw their task through to the end no matter how difficult, would not shirk from their assigned duty; they were dependable, you could count on them.

Like Major Thompson, our squadron commander, our lead pilot, twenty-four years old, one of the youngest majors in the Air Corps.

He emphasized how he was a professional, he said he'd do anything he was ever called upon to do to advance his career. He said he wouldn't let anybody or anything get in the way.

Then there was the other kind, the maggots, the parasites, the cockroaches, the weasels, the sneaky fuck-offs. The kind that hide from responsibility, avoid it at every possible opportunity, when stuck with it, do the absolute minimum, used all their energy, all their cleverness to avoid doing anything. Now, that's the kind of guy I am.

I hate responsibility, I can't stand it, I despise it. Responsibility scares me, makes me nervous. Fuck responsibility.

See, what happened, the American bombers were knocking the hell out of Germany and one strategy the Germans had was to sail suicide commandos down the Adriatic and blow up our bombers on the ground.

It was happening; so our air bases had to be guarded by tough infantry guys.

The way it worked was that the air base commander would assign an O.D., Officer of the Day, to be in charge of the guards and the security of the whole base for twenty-four hours. It was a pain-in-the-butt job. I was successful in avoiding it for months, until I got nailed with it.

The day I was stuck with it, I went to the control tower and relieved the O.D. that was on duty. He handed me the pistol and introduced me to Sergeant Lemoyne Smith, U.S. Infantry. We looked at each other, sizing each other up. Sergeant Smith was, I guessed, about ten years older than me and about a foot taller, a big guy. He was black and spoke with a New York accent.

I said, "Look, Sergeant, I'm a flying officer and I don't know shit about infantry troops or guarding a base; what do you suggest I do now?" He smiled and said, "Don't worry, Lieutenant, I'll take care of everything," and he did.

After Sergeant Smith posted all the guards, I asked, "Whata we do now?" He said, "We just sit here all night long and stay awake. Let's play some gin rummy."

Sergeant Smith was a very bright guy. He'd played football for Columbia and had a master's degree in business administration.

Time just dragged on and on. Finally daybreak, a guy brought over some breakfast. Then lunch. Then it was twilight; my stint was almost over.

Sergeant Smith and I were outside when a low-flying plane came out of the eastern horizon, the Adriatic side. Smith said, "Lieutenant, what kind of plane is that?" I looked at it and said, "Well, it's not one of ours, maybe it's British." The plane had banked around, getting ready to land. Then I recog-

all piled in and two hours later, we were back at our home base and debriefed, and in our tents.

The next day, Danny showed up in my tent, calm, pleasant, handsome as ever, looking almost beatific. He said, "Rockets, I gotta talk to you." And then he told me that when the war was over he was gonna become a priest. I asked, "Why?" He answered, "Rockets, you know why we made it back. It was because I made a deal with God; Rockets, when Bobby got killed and all of his blood was spattered all over me, I was never so scared in my life. Rockets, I shit in my pants. And I started to pray. I took a vow. I said, 'God, just get me through this and I'll do whatever you want, God. Honest, I will.' And God answered me, he really did. God said, 'Danny, what do you like to do the most?' and I answered, 'God, I just love fucking.' And God said, 'Danny, you promise to give up fucking and I'll get you through' and I said, 'I will, God, I promise' and then God said, 'I'll get you through' and he did."

Then Danny said, "Rockets, I can't help it, I'm just horny all the time. Pussy. It's all I ever think about. It's the way I am. It's a real sacrifice, and I might as well become a priest," and he started sobbing.

I thought for a bit and said, "Why don't you talk it over with Father Pietro."

Danny said, "Good idea" and he did.

Father Pietro asked, "How many missions you got left to fly, Danny?"

Danny said, "Fourteen."

Father Pietro thought and then he said, "Danny, my son, a deal is a deal, a promise is a promise, a vow is a vow, so you should abstain for the next fourteen missions. It won't kill you. But you can still see Angelina and accept a limited amount of her comfort."

Well, that's what old Danny Callahan did. He told me that he explained the predicament to Angelina, no more fucking for the next fourteen missions. So they did other stuff. She'd go down on him, he'd go down on her. He said they 69'd a lot and even 72'd. I never figured out what a 72 was. I asked Danny, he just smiled and said it was just three more than a 69. He said they just had a hell of a time. But no fucking.

I went to the Bombardment Group reunion in Las Vegas last year. I ran across Danny. Didn't recognize him. Just saw his name tag. I told him what I had learned from him that day. That often emotions are just contagious. You can catch emotions like you can catch the flu or a bad cold. Like, say, fear, or hate, or love, or bravery can spread like germs in the air you breathe. And even if you're scared to death and you just fake being brave, everybody around you will be brave.

Danny looked at me like he was trying hard to remember, then he blinked his eyes and said, "I'm sorry, who are you? What the hell are you talking about." Then he ambled away, using his aluminum walker, mumbling to himself.

parked, there was a light on inside. I knocked, no answer. Knocked again, no answer. So I walked all around where I knew the Colonel's office was, and put my face against one of the high windows.

"Aw, Jesus."

You know, there are certain images that are so strong, so powerful that you never forget them. It's just tattooed in your brain.

Well, there was our Colonel and Major Thompson, our squadron commander, stretched out on this bed, all naked. White bodies, all asses and elbows, this jerky motion, they were in that head-to-toe 69 position, they were sucking each other off.

I was stunned, mesmerized, knew I was watching something I shouldn't be watching, maybe I just watched too long. Hell, I didn't even know guys did that sort of stuff. Jesus. I had never even been to a porno movie and here it was—live.

Then I thought, "Hell, now what'll I do? If I went back to the control tower, that would take another twenty minutes. Then what?"

I went back in the jeep and decided that I'd just start all over. Put the jeep in reverse and went back about fifty yards, then I revved up the engine and drove fast as hell toward the headquarters, honking my horn and slamming on the brakes as I skidded on the gravel. I honked and then jumped out and started banging on the front door, yelling this time.

"Colonel, Colonel." Making a racket like I had just gotten there.

I heard the Colonel yell, "Just a minute," and then after a bit, he opened the door.

He was dressed in a silk paisley bathrobe and soft Italian leather slippers. He had a mess of shaving cream on his flushed face that looked like it had been thrown on.

He looked at me and then squinted, "Oh, it's you. What do you want? And it better be good."

I explained to him that a German Messerschmidt had landed and given itself up and that the phone was dead and that I thought it was unusual enough that he should know right away.

The Colonel shouted, "What, Oh! You did the right thing, son. Let me get some clothes on and drive me down there right now. Wait here."

I waited, the Colonel disappeared inside, and I peeked in. The first thing I saw was the phone. It was off the hook, no wonder I couldn't get through.

I drove him down there. I explained what happened between Sergeant Smith and the German pilot. Didn't bother him one bit.

The Colonel was talking about something, but I wasn't listening.

All this stuff was going on in my head.

I was starting to feel pissed off. Here I had been up all night, all concerned about guarding our air base, and sort of feeling guilty that I wasn't responsible enough, and all the while the Colonel and the Major were having just a hell of a time sucking each other off, and probably having champagne and caviar to boot.

Yeah, I know rank has its privileges, but even so it still pissed me off.

And I don't deny that any kind of sex was of great interest to me anyway.

Now everyone sort of knew that the Colonel was kind of queer, but I was really surprised about Major Thompson, our squadron commander.

I wondered was he just paying the price to get a promotion, or was he just horny and this was a one-night stand, or did he and the Colonel really have a thing for each other.

As soon as we got to the control tower, the Colonel said, "Lieutenant, I'll relieve you now, I'll take command. Go get Captain Jones and have him bring that photographer of his."

Captain Jones was the group's public relations officer. I found him at the Officers' Club, told him what happened and he got excited and said this is "Big—big—really big."

Two days later, a story came out in the *Stars and*

Stripes with photographs of the Colonel in front of the German Messerschmidt, alluding to the great job the Colonel had done in capturing it. I could identify that easy. It was just wonderful image-making bullshit.

By a week later, things had calmed down, the excitement was gone, we were back to routine stuff.

You know fifty years have passed, but those guys are just stuck in my brain. After the war, the Colonel stayed in the Army and became a general and ulti-mately retired and lives in Florida. Sergeant Smith became a stockbroker in New York City and is very rich. Major Thompson is buried in the American Air Force Cemetery in Florence, Italy. I don't know what happened to the German pilot.

And me, I'm just trying to follow Maria's advice.

She used to tell me over and over. Learn to live in harmony, walk in beauty, live a graceful life.

That's what I'm trying to do.

And as for responsibility. I hate responsibility, I can't stand it. It makes me nervous. To hell with responsibility.

MAJOR BRADLEY DUNCAN BELCHORE III

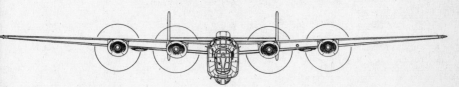

THIS WAS THE DEAL. THE ARMY needed a bunch of flying officers in a hurry, so they started a program called Aviation Cadets. But for the first six weeks they ran you through a ritual called "boot camp" as sort of a weeding-out process.

That's where I met Bradley Duncan Belchore the Third. We were in the same barracks and for some reason, Brad sort of sought me out, we became friends. He was an elegant, handsome guy from San Francisco. His family was wealthy, Belchore Salt. He had gone to a private prep school, the Whitney School in La Jolla, and then to Stanford. He got a B.A. in political science and then he joined the Army Air Corps.

The bugle would go off at four in the morning. You had to jump into your clothes and run outside and line up for roll call, then you had to double-time over to the mess hall for breakfast, double-time back to the barracks, shave, go to the toilet, make up your bunk, shine your shoes, and be ready for inspection at 4:30 A.M. After inspection, the routine would start, marching, all kinds of endurance stuff to try and either build you up or wear you down. Like you weren't allowed to walk anywhere, you had to run. Everything was double time.

It was during that four to four-thirty time where Brad had these little problems.

There'd be about three barracks full of guys, that's one hundred twenty guys to use the latrine at that short time. There were twenty toilets, all in a row, no toilet partitions, no toilet seats, so there'd be twenty guys all shitting at once. You had to take your turn. And you know guys would be grunting and wiping and flushing and there would be all these, you know, all these natural noises that you'd just as soon not have to share with anyone else. If you were at all sensitive, well, it just wasn't an aesthetic experience. And then there were twenty sinks and you had to shave and brush your teeth and reach over the guy in front of you. It was a madhouse, but you didn't have a choice. You didn't have a free minute until nine o'clock that night, when it was lights out.

Well, Brad couldn't handle it. He told me, "I'm not an animal." He couldn't shit on demand and if there wasn't any hot water, he wouldn't shave. He worked out this system of waking himself up at 2 A.M. when the guards were asleep, dressing, and then walking about one hundred yards to the latrine building; but it was February in Lincoln, Nebraska, and it was snowing and freezing outside. It was difficult, but that's what he did. He said, "I just need my privacy."

And he told me about his family. His great grandparents were from Bavaria and they had heard about "gold" in California in 1849. They bought tickets, got on a boat, and sailed to the new world. After landing, they were there a month before they found out they were in Scotland, not America. They stayed in Scotland for ten years till they saved enough money to start out again. Children were born and were given those Scottish names like Bruce, Duncan, Bradley. In San Francisco, they went into the only business they knew— salt. Salt for people and salt for livestock. Later, things grew and they got into the Death Valley borax business. In World War I, their chemicals were in demand and they became wealthy; by World War II, they were super rich and Brad was the heir to the Belchore Kingdom.

We had all been tipped off by our drill sergeant what would happen and what we were supposed to do and say.

We would be brought into a waiting room, in groups of ten guys. We were to take off our clothes, down to our shorts, and then wait to be called in, one by one, then salute the examining officers, when asked what we wanted to be, the correct answer was, "A pilot, sir." "Why?" "Because I want to fly and fight for my country, sir." And that was it. That's what we were supposed to say.

See, the U.S. Government was going to invest a lot of time and money in us and they wanted as few fuck-ups as possible.

And sure enough, maybe one out of every ten guys would blow up, saluting in his underwear and all and raise hell about being kept waiting so long and to hell with you, sir, and, of course, that's what the psychiatrist was looking for and he'd call it "emotional instability" and that guy would then be pulled out of the Air Corps and sent to the infantry.

And with the psychiatrist there was an evaluation and assignment officer.

When it came to my turn, I did all the right things

and gave all the right answers, and the assignment officer looked over my records and asked if I would accept navigation school, instead of pilot school. I asked if I had any other options and he said paratroops or infantry. So I said navigation school sounded fine, saluted, and left.

Well, back at the barracks we were all excited and talked about what we got. Brad was one of us that got navigation school and he was livid.

Brad, by his nature, was a pretty cool guy. He wasn't given to flying off the handle. He could absorb a lot of abuse. Like, probably the best name to have in the Army was Smith or Jones. If your name was unusual, when it was called out at reveille roll call, you got laughed at a lot.

There was one guy whose name was Zuchov. The sergeant called out his name as Suck-off, another poor guy's name was Fuchs, it became Fucks, Rudidick became Rub Your Dick, Harden became Hard-on. When it came to Belchore, the sergeant would burp first and then call Belcher and everyone laughed.

Brad was used to that, he would take it. But he wouldn't take his assignment to navigation school.

See, not only was he already a licensed pilot, but he was rated in instruments, twin engines, and was an instructor as well.

He said it was "bullshit," and he wouldn't take it. He said he'd show 'em and that night he called his dad, and explained the situation, and his dad called the senior senator from California, and the senator called somebody else and the next day, Brad was called in by a major who reviewed his credentials and he was sent to Randolph Field which was the best Air Corps flying school in the country and was asked whether he preferred a single- or multiple-engine plane and he said multiple and he was put in a B–24 bomber training base and that was that.

I said, "Brad, you've got a lot of guts."

He said, "You haven't seen half of it yet."

He accused me of being an Uncle Tom, afraid to make waves, afraid of confrontation, a coward, and you know, he was right.

Shit. I wasn't about ready to fight the Army.

When I was in navigation school, we would fly all over the United States on training missions and we'd land at different bases. I'd often see other cadets that I'd met at one place or another. I'd ask about Brad and I'd pick up just bits and pieces of news.

One was that he was already commissioned a second lieutenant and had his own crew and was on his way overseas. I felt sort of retarded, I was still a cadet.

Another guy told me that when Brad was commissioned, well, all the other guys simply went to
the PX and bought their "off the rack" officer's uniform for seventy-five bucks—but not Brad; he had
tailors brought in from New York from a company
called Abercrombie & Fitch to fit his uniform that
they had made for him. And that he looked real
sharp.

When I got to Italy, at Orto Vecchio, damn if
Brad wasn't in the same group I was assigned to.

I was a second lieutenant and he was already a
captain.

He was already sort of a mythical character. He
had already flown thirty-five missions. His crew
loved him. He made a flat-out guarantee to his
crew that nobody would get hurt and everybody
would make it back safe and of course it was bullshit, but the guys on the crew believed him and,
sure enough, in some of their narrow escapes, it
was as if it was Brad's will that got them back. He
was legendary.

Brad's crew was hand-picked. When not flying
combat, they were practicing aerial gunnery and
flying. He knew his ground crew. They were the
best. No engine troubles for Brad. They were all
sharp and they just adored Captain Bradley Duncan Belchore. They were loyal. He was working his

magic. Good-looking, sharp, funny, twinkle in his eye, a natural leader.

Every month, we got paid and Brad would take his dough and split it up among the enlisted men on the crew as well as the ground crew, the maintenance guys.

When any of his guys needed extra dough for trouble back home, Brad would put it up. He was straight about it. "Look," he said, "I'm rich. I've got more money than I need already, and if I can help out a guy that I care about, what the hell."

And he looked plenty dashing. He always wore those handmade soft leather boots from Argentina.

He'd tell me, "All it takes is guts, imagination, and a little luck and you can have everything, power. Real power.

"I'm going to get every medal I can. I'm setting my sights for politics and after this war is over, you'll see me. I'm going to be the youngest colonel the Air Corps ever had. Then the youngest senator.

"It's posturing, perception, and bullshit. See how Patton and MacArthur get away with it? Well, so is Bradley Duncan Belchore. Just watch me."

I thought, Brad you're just fucken nuts. And then I thought, maybe he's not. I didn't know.

Shit, he was planning on being the next king. That was how it was, he was the prince and I was the peasant. Then I wondered if he was just the court jester or maybe we all were. I was getting confused.

Brad and his crew finished their fifty missions and Brad threw a party. He paid for drinks and somehow or other produced porno movies, arranged for some awfully raunchy-looking females to come in from Foggia and then in the middle of the celebration said a tearful goodbye to his crew who were going home and announced he had signed up for another fifty.

By now, he was twenty-four years old. He had been awarded the Distinguished Flying Cross, the Air Medal with ten clusters, the Presidential Unit Citation. At dress-up occasions, his fruit salad medals looked dazzling on his chest. He looked good.

I thought, "Fifty more missions? Brad, you're asking for it."

Brad kept the same plane, got a new crew, was promoted to major, and was moved to Group.

He was by far now the most experienced pilot on the base and most often flew the lead plane of the whole damn group.

He and our colonel, our group commander, got to be good friends and he asked Brad to move in with him.

The Colonel's Group headquarters was a modest two-story stone farmhouse on top of this hill, but next to the tents that we lived in, it looked huge.

It was shabby-looking until Brad moved in with the Colonel.

Brad brought in some beautiful Oriental rugs that he flew in from Morocco. He brought in Talivera dishes from Sicily and Naples, found some great old Italian oil paintings, lined up elegant wines and good whiskey, hired a good Italian chef, some fine Italian antiques, and the old farmhouse started looking pretty good. Brad had this ongoing trouble with toilets. There were no flush toilets on the base, just very smelly outhouses.

Brad found some Italian tile setters and put them to work; he had them build an oversized tile bathtub, and a large elegant tile shower. He found a genuine sit-down flush toilet and had that installed. He rigged up a rooftop reservoir of fifty-gallon metal drums full of water heated by one-hundred-octane aviation gasoline so they always had hot water. That bathroom was the marvel of the base.

Pretty soon, the word got out that our colonel threw pretty good parties, and big shots, colonels and generals, started showing up regularly.

Brad knew how to play the game.

It became obvious that Brad and our colonel were very chummy.

Brad was teaching the Colonel certain social and political skills that were required if he was to become a general.

All kinds of politics were brewing on the base. Our colonel, as group commander, had four squadrons under his command. Each squadron was commanded by a major and all happened to be West Point grads. They were professional soldiers and they didn't like our colonel, him just being VMI, Virginia Military Institute. And they knew he was bucking for general and they knew he would appoint his heir apparent and they were all ambitious and they didn't want Belchore, the glamour boy. They hated Belchore and were afraid of him.

Now Brad sensed this and the way he handled it was to work out a deal with the Colonel. The Colonel would pass the baton to Brad if Brad's dad pulled the right strings back in Washington and then both would have what they wanted. The Colonel would be general and then Brad would be the youngest colonel in the Air Corps.

But Brad knew he had to have the respect of everyone on the base and the way he got that was by flying lead on all the tough targets.

See, the tradition was that the squadron and group commander, the majors and the Colonel, could select what missions they would fly on and they usually selected those easy milk-run, chicken-shit targets.

Like bombing Sofia, Bulgaria, or the Tatoi Airdrome in Athens, Greece, or those little undefended railroad towns that were in front of the advancing Russian armies. Shit, they were nothing, but to go bomb the Krupp plant in Schwienfort, or the Messerschmidt factory in Ausberg, or the BMW factory in Munich, the Germans threw everything at you and we paid for it, you came back, if you came back at all, bloody and terrified and wounded and hurt.

The commanders wouldn't go on these missions, but Brad did and everybody knew it. He was paying his dues.

On August 23, 1944, the target was Ausberg again, and Brad was flying lead and it was like the whole German Air Force came after us. But we flew a tight box and although we lost six bombers that day, almost a whole squadron, we shot down fifteen fighters, a record, Brad's plane got three; and he was awarded the Silver Star. In the Army, the Silver Star is not a small trinket, it's big potatoes, but Brad wouldn't take it unless everybody on his crew got one too and that's what happened and the word spread all over the base and even to other bases and Brad became a guy deeply respected, and the West Point majors hated him even more.

Then a month later, in September, Brad was flying lead again, we had just bombed the BMW plant in Munich and we turned and were going home to Italy when our group came across a lone B–24. One engine was out, one engine was smoking, all alone and going in a screwy direction, going east instead of south. The lone bomber had a big hole in its side and obviously was in trouble. Brad radioed the plane, slowed our whole group down, and got that lone plane to join us, otherwise it was just certain to be picked off by German fighters. But the lone plane couldn't hold altitude, so Brad brought the whole group down to join it. We just skimmed over the Alps.

It was just like in the Bible, do you risk the whole flock to go after one stray sheep? That's what happened. When we got back, the other majors just started raising hell about what Brad had done and they wanted him court-martialed and demoted because it was against all Air Corps protocol.

There was a lot of yapping about it; and then the word got out that the pilot of that stray bomber was a general's son. Oh, that changed the whole picture. The general was an Air Corps big shot in Washington,

D.C. Our colonel received a telegram from the general complimenting him, personally thanking him, awarding the whole group a Presidential Unit Citation and is there anything else I can do for you, Colonel?

The West Point majors muttered that lucky son-of-a-bitch Belchore, but had to shut up.

The ground crew rumor mill on the base was working overtime. The word was our colonel and Brad were getting awfully close, some soldier walked in on them when they were having a pillow fight and they were wearing bright red bikini underwear, another guy swore he saw them dancing together.

One day, Brad came to my tent and asked me to drive to Foggia with him. We went to a good restaurant there and we had a few drinks and loosened up and then he asked flat out what the gossip was in the trenches.

I said, "Brad, you don't want to know."

Well, he wouldn't let me up. He insisted on every detail.

I told him and he was pretty calm.

He said, "That's not too bad. I just got to do a little damage repair, that's all."

He said nobody had any real proof of anything and he could cool it with the Colonel for a while.

We started getting drunk. I finally blurted out.

"God, Brad, how do you stand the Colonel anyway—he's such a jerk."

Brad answered, "You're so damn dumb. You're
pathetic.

"Don't you know the only thing in life that's
worth a damn is power and you get it any way you
can. You buy it, you fight for it, you fuck for it, you
do anything and once you got it, you hang on to it.
You try and get all you can. It's life itself, it's everything, everything, there's nothing else.

"And what's so bad about screwing him. Hell,
he's cleaner and sweeter than those smelly, hairy
whores in Foggia.

"And whose business is it anyway, who I sleep
with.

"And all this fuss about sex. All sex is is just a little bit of moisture."

Brad said, "Oh, if you're so smart, do you know
what you want out of life? What do you want?"

I had been raised by a Mexican Indian woman,
Maria.

I said what she had taught me, "If I had my way,
I want to walk in beauty. I want to live the graceful
life. I want to live in harmony."

"Jesus," he exploded. "You're such an asshole.
You don't know anything. You make me sick, that's
bullshit. Do you really believe that shit?"

I said, "Yeah, I guess so." I wasn't so sure any-more.

In mid-October, on Brad's seventy-second bomb-ing mission, the target was the BMW plant at Munich again; and Brad got it.

After the war, I was a student at Berkeley. I had gotten married and one day I called up Brad's folks and they got all excited. Apparently Brad had writ-ten a lot about me. They invited us over for dinner.

We showed up. They lived in this huge mansion in Pacific Heights overlooking the Golden Gate Bridge. They had butlers and maids and all.

At dinner, they served these shrimp that had been split in two. I'd never seen them before, my wife explained the shrimp had been butterflied.

I thought, "Jesus, that's just what happened to Brad. He'd been butterflied, just ripped in two."

By the time his plane landed, he had been dead for hours. His crew cried.

Bradley Duncan Belchore the Third was buried at the Army Air Corps Cemetery at Bari, Italy.

I said the memorial prayer.

OLD MAN IN A BASEBALL CAP

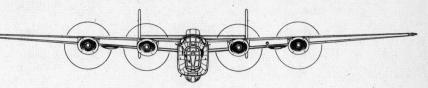

By MID-FEBRUARY OF 1945, I HAD forty-five missions to my credit, just five more to go and my tour of duty would be over and I could go home.

On that morning, we went out and bombed an oil refinery in northern Czechoslovakia and over the target, one of our engines was shot out and then as we turned to go home, another engine started smoking badly and after a bit it quit on us. And from then on, we couldn't maintain altitude. We were losing about five hundred feet a minute.

And we had to start bailing out, and we were over enemy territory.

Now, when you're over enemy territory, you bail

out individually, that way if you're captured it won't be all in one bunch.

When it came my turn, I jumped at six thousand feet. We were over Yugoslavia. When you jump, we were told to count to ten, so you clear the plane, and then pull the ripcord. I counted to three. And these were old-fashioned parachutes and they opened up with a terrible jerk. It was like getting hit between your legs with a baseball bat.

And then about fifty to sixty feet from the ground, the damn parachute collapses and you fall like a rock.

I landed on this farm. There was a bit of snow on the ground. The tilled land was frozen solid; it was like concrete. I hit hard.

I was dazed. I was seeing double. There was blood all over me. Some guy was standing over me screaming, "Ruski. Ruski." A farmer. I yelled back, "No, I'm Americanski."

He dragged me to a stone hut near the edge of this field where he had stored some of his tools. I lay there for two days. I was hurt pretty bad. My jaw was fractured. Three ribs were cracked. It hurt like hell every time I breathed. And my hip and leg were banged up and badly bruised.

At night, the farmer came in and left me some black bread and a bottle of this stuff. Slivovitz.

In the evening of the second day, these three people came into the hut. Two tough-looking guys and a woman. They were wearing heavy woolen British infantry uniforms with red stars sewn on. They were Yugoslavian partisans. They stared at me. They talked. I couldn't understand what. It sounded like they were negotiating. Finally, the woman spoke, "Me, Maruska Vlahovitch. I speak English. I walk you back to Italy."

I said, "You're joking."

"No."

"How long will it take?"

"It about four hundred kilometers to Vis, Tito's headquarters. We walk forty kilometer a day, it take ten day. We walk twenty kilometer a day, it take twenty day. But we only walk at night. I think it take thirty day."

Then she winked and said, "That is if Germans don't find us first and shoot us."

She smiled at me, "You got no choice, you can't stay here. You be all right with me. I strong."

Then the two men said something and left.

That was that. I looked at Maruska Vlahovitch and she looked at me. She said, "What you name?" I said, "Fred." She said, "Me, Maruska." An hour later, it was dark and we started walking. Each of us had a loaf of black bread, a bottle of slivovitz, and a woolen blanket. I did better than I thought I would. It wasn't too bad. It was cold and dark. She explained that was good. The Germans hated cold and were afraid of the dark. The partisans owned Yugoslavia at night and hid by day. That's what we were going to do. Walk all night and sleep all day.

We walked all night. At about dawn, we were near this village. Maruska said, "Wait here," and went to one of the huts. She called out, then said something like, "Americanski," and then called, "Frat, come here," and we went into the little barn, ate some bread and drank some slivovitz. Then Maruska and I stretched out and went to sleep. At dusk, a guy came and gave us a bowl of potato soup, some bread and slivovitz. Maruska and I hardly talked, we were strangers and she was just doing her job. She would give me commands though, "Frat, go. Frat, this way. Frat, stop. Frat, drink."

Maruska had a good idea of who she could approach for food. All anybody had was a little cheese that looked like loose watery cottage cheese, and a bit of black bread. There seemed however to be a surplus of slivovitz, the plum brandy. And that suited me just fine, it numbed my senses, eased my pain, and Maruska and I were semidrunk most of the time. Really, I was drunk all of the time.

We walked all that next night. At dawn, Maruska pulled us off this little dirt road we were on. We found a spot in the woods. There was still a bit of snow on the ground. It was cold. We spread out a blanket on the ground, lay down on it, and then pulled a blanket over us. We huddled next to each other for body heat and went to sleep, but it was awkward.

Maruska had been warned at one of the villages that a small detachment of about fifty S.S. German troops were about two days' march ahead of us, and we should stay off the main roads and stick to the smaller trails.

At the end of the third night of walking, we pulled off the trail near a creek, spread out our blankets, drank some slivovitz, went to the creek, washed our faces in cold water, had a drink of water in our cupped hands, and at daybreak, fell asleep in the woods.

About three or four hours later, Maruska woke up and just started vomiting and vomiting. She couldn't stop and then she clutched her stomach and was just doubled up in pain and then started having a violent diarrhea attack. She was in awful pain, was shivering, had a fever, just sick as a dog. Her face was bright pink, unnatural. She would just lie there, I didn't have anything to wipe her

with, we had no toilet paper or anything. I took off
one of my socks and wiped her butt with that and
saw that her ass had turned bright, bright red, like
a baboon. And that her shit, the shit that was coming out of her was like black ink, jet black. And
that scared me and I thought, "This gal is dying."
Then I had these thoughts, "If she dies, what'll I
do? Bury her or just leave her? And then what?
What do I do next? Where should I go?" I felt help-
less. I didn't know what to do. Then about an hour
later, I started vomiting and vomiting and the
damn diarrhea, not as bad as Maruska, but bad
enough. I figured, and as it turned out correctly,
that we had been poisoned and the best thing to do
was to try and get the poison out of our bodies so I
forced slivovitz down her throat and drank a lot
myself. We were both feverish and sweating and
we just lay there together all the rest of the day
and part of the night.

At about midnight, Maruska woke me up and
said, "Frat, we go"—God she was tough—and we
started to walk.

"It bad water," she explained. She had been
warned of a rumor that the S.S. were dumping poison into creeks and wells, a mustard gas by-product,
to drive out the population. The rumor was true.
We walked the rest of the night.

At dawn, we came to this little village—God, what a sight. There were ten little huts all burned down, a couple of huts were still smoldering, and there in the middle of this little street was a gallows with eight bodies hanging from it. Six men and two women, hanging there like little birds. And they had hung the women upside down.

Maruska cursed the Germans. Said, "Nothing we can do," and frantically started searching the burned huts for food. We found six potatoes, a loaf of black bread, and a bottle of slivovitz, and it was already light and we hurried into the woods to hide. We were frightened. We figured the S.S. wasn't too far away.

We found a spot, spread out our blankets, drank some slivovitz, ate some bread, and went to sleep.

About four hours later, Maruska woke me. It must have been about noon. Maruska said, "Frat, you think I no beautiful?"

I said, "What the hell are you talking about?"

She said, "Me, Frat, me. You think I no beautiful? Frat, you saw those people hanging in village. Frat, if S.S. catch me, that what they do to me. If they catch you, they will hit you and make prisoner. Me, they will hang. Frat, I virgin. I don't want to die virgin. Why you don't put your hand on my siski? Why don't you put your hooey in my pizka?

Why we don't make fig-fig? Frat, you think I no beautiful?" and she started to weep and sob uncontrollably.

Well, I was taken aback. I knew she was risking her life for me.

I said, "Look, Maruska, first I think you're very beautiful, and the Germans aren't going to catch you. I won't let them." I wasn't sure about that.

"And Maruska, I've got this problem. My jaw is fractured, these ribs are cracked, it hurts like hell every time I breathe. My hip and leg are bruised and banged up and we're both covered with shit from the waist down. God, we both smell like a sewer."

She looked at me and said, "Just excuses. Any Serb man make fig-fig with me by now."

Then she said, "Frat, you really think I beautiful?"

I said, "Like a movie star."

I put my arm around Maruska and we closed our eyes. She had stopped weeping and was calmer. We lay back and tried to go to sleep.

I thought she did have a point though. She was right, it was excuses. Look, she wasn't beautiful, she was tough and wiry, dark, furtive, sharp-featured, sort of hatchet-faced; no, she kind of looked more like a ferret. But she was right. I had been thinking about making a pass at her, but it was now six days since I had bailed out but in all those six days, I hadn't had a hard-on once.

Now you know, when you're twenty years old and you're a male, you have a hard-on damn near all the time. You certainly wake up with one and then episodically have one the rest of the day and night. That's just the way it is.

Well, what happened to me? I was worried. Had that damn parachute jump damaged my balls?

We walked all the rest of the next night. At daybreak we came near this village. These eight tough-looking guys, four on each side of the trails, surrounded us. They had these British burp guns, they're really called Sten machine guns, when you pull the trigger, they go burp. That's why they call them burp guns. Maruska explained who we were and they got excited and I kept hearing Americanski, Americanski. And they walked us into the village and then they started just raising hell. They gave us some hot soup, and bread and slivovitz, and they killed a lamb and started to roast it on a spit over an open fire.

They poured a bunch of warmed water in a tub in one of the huts. Maruska and I took a hot bath with real soap. They took our shitty clothes and washed them. While they were drying, a guy came in and gave me a shave. Maruska and I lay on a real bed and went sound asleep. Oh, it felt just great.

FRED ROCHLIN

At about five o'clock, they brought us our clean clothes. We got dressed and went outside and there was all this excitement going on. And the head man of the village came and gave a talk and I kept hearing Americanski and then we all started to walk toward this hut on the edge of the village. Maruska said, "Frat, I know you. You not a hard man. They will give you a gun and you must do it even if against your nature, you must, or they kill you and me." I didn't like the way that sounded. It gave me the creeps.

And the head man came to this hut and pulled hard on the chain and from under the hut crawled these three young blond, blue-eyed German boys. They had S.S. insignias on their uniforms. They were filthy, covered with shit, had been terribly beat up, maybe they were seventeen or eighteen years old.

The head man stood them against the wall, their arms and legs chained, and they had a chain around their necks like a beaten dog.

The head man smiled and said something and I heard the word Americanski and they handed me the burp gun.

Maruska quietly said, "Frat, do it. You must do it." And all the people started yelling, "Chata, chata, chata, chata."

Jesus Christ. I didn't want to shoot anybody. Maruska kept hissing, "Do it. Do it. You must do it."

100

I did. I pulled the trigger and the gun went burp and I sawed those three boys in half. I was half numb. I saw one boy look at me and he cried, "Mutter." I saw the blood spurt out.

All the villagers shouted with enjoyment. I thought how easy the whole thing was. It didn't take any courage, you just pulled a trigger.

After, we all sat down to a meal of roast lamb and potatoes and vodka. Then the villagers loaded us down with as much food and drink as we could carry and at nightfall we started on the trail.

At dawn, we spread out our blankets in the forest and lay down to go to sleep. I stretched out and to my surprise I had a hard-on. I reached over and, for the first time, gently kissed Maruska. Gently, because my jaw still ached like all hell. I unbuttoned her jacket and put my hand on her siski. A little later, we were naked under the scratchy woolen blankets and I put my hooey in her pizka and we made fig-fig in the Serbian forest.

And so a pattern started that was to go on for the rest of our journey. We would walk all night and at daybreak would find a spot in the forest, spread our blankets, eat some bread, drink slivovitz, make fig-fig, and go to sleep. We would wake up at dusk, make fig-fig again, roll up our blankets, eat some bread, drink slivovitz, and start walking. That's what we did.

But on the second day after our initial love-making, or screwing or fig-figging, whatever you want to call it, a terrible itching developed in my pubic hair and I examined that and hell! Lice. I had gotten crabs from Maruska. And then the next day, I went to take a leak and it was like my hooey was on fire and it started to drip stuff.

I said, "Maruska, you son of a bitch. You've given me a dose of clap. Gonorrhea and crabs. God damn it. What was all this bullshit about you being a virgin?"

She said, "Don't get so excited, everybody got it in Yugoslavia. Is nothing to get excited about. And I am virgin. I still virgin. I virgin in my head and in my heart, just not my body.

"And Germans probably find us and kill us anyway.

"And I still don't want to die virgin.

"Frat, you still think I beautiful?"

I just grumbled something and we both just sat down. We pulled down our pants and started picking out the lice. She started crying again.

The next twenty-one days and nights were relatively uneventful. The farther south and west we got, the friendlier were the natives. We still had to walk at night and hide out in the day, but there were obviously more partisans around, a bit more food, fewer German soldiers in the countryside.

As things got easier for us, my leg and jaw and ribs were healing. Things were looking up, but Maruska seemed to get more nervous and testy. And she seemed to be on my back a lot of the time. She'd say, "Frat, you drink too much, you drunk all the time." Well, that was true enough. Then, "Frat, walk faster," or "Frat, walk slower." Every night just before we started to walk, she'd say, "Frat, tie your shoelaces." As if I had no brain at all. No matter what, she had to have the last word. She could be a motor mouth. She could talk endlessly about nothing. She would repeat herself on and on. I heard the story of her childhood at least twice a day, every day.

She was saving my life, but she was getting under my skin. She was driving me nuts, not on purpose, just her way. She seemed happiest when she was harping about something or criticizing me for some infraction or lecturing or telling me how right she was and how wrong I was. Then, she'd start the inquisition. "Why did you do this? Why did you do that? What's wrong with you? You're not telling the truth? Why don't you talk more?" and on and on until I'd explode and say, "Shut the fuck up. For Christ sake, shut up." She'd say, "Don't talk about Christ like that."

My explosions would slow her down about ten minutes and then the mouth would start up again.

One night, we had slept, woke up, fig-figged, eaten bread and slivovitz, and rolled up our blanket. It was nighttime and we hit the trail, but she started in the wrong direction. I said, "Maruska, it's this way, not that way." She got furious that I'd question her. She said, "Frat, you full shit. This is way. I know which way to go." I said, "Bullshit, that's the North fucken Star, right there. See it, and I'm not going to walk fucken north toward Germany. I'm going to walk south. Get it?" She looked puzzled. "What stars?"

I said, "God damn it, right there, that's the North Star. That way is north and we want to go south." She was adamant. "You shit, you know nothing. Don't talk to me about stars. This is the way," and she started walking north.

I yelled, "Well, goodbye Maruska and fuck you. I'll be god-damned if I'm going to walk one step in the wrong direction."

And she kept walking. I just stood there not sure what to do.

After about five minutes, she walked back and she said, "Okay, they command me no matter what, I must stay with you. But you big asshole. You walk in wrong direction." At daybreak, we came across a shepherd and she talked to him and it was obvious that we were going in the right direction, going south. She looked at me, then spit at me and then smiled. Then asked, "How you know about stars? You full shit."

I said, "Maruska, sometimes you just gotta shut up. You're driving me fucken nuts."

She said, "Frat, I know, I can see it in your eyes. Why when things are very hard, difficult, we no fight, we get along fine? Why when things get easier, we fight all the time?"

I looked at Maruska, "Jesus, I don't know."

She said, "Frat, don't use Jesus name like that."

I said, "Oh, Christ." And we kept walking.

On the morning of the twenty-eighth day, I had calculated that we were only about thirty kilometers from Dubrovnik. Just a two-and-a-half-day hike. We were getting closer and closer.

That morning, we walked a little later than usual. Maybe we had gotten careless and this tough-looking guy stepped on the trail and said something to Maruska in sort of a challenging tone. I couldn't understand. She answered back, and then a loud argument broke out. I didn't know what it was about, but then he pushed her down hard and kicked her while she was down and looked at me and sort of smiled and signaled with his hand that I was to come with him. I had a walking stick, sort of a staff. I'm not a violent type, but I got furious and whacked the guy as hard as I could with the stick on the side of his head. He dropped and I whacked him again and again. I went a little nuts. Maruska got up and said, "Frat, stop." Then she bent down and sort of helped the guy up, gave him a shot of slivovitz and said something to him. And we left him sitting on the ground.

I said, "Maruska, what was that about?"

She said, "Nothing."

I said, "Bullshit, what was it about?"

She said, "Better you don't know."

I finally yelled, "God damn it, what the fuck was that about?"

She said, "All right, it is simple. There is a twenty-five-hundred-dollar reward in gold for anybody who brings American flyer to Vis alive. That's all. He just wanted you for reward. He say woman don't need reward. He say man need reward, so I say fuck you and he push me down."

I said, "Oh."

She said, "Frat, I like the way you fight for me. I like way you want to protect me. Frat, I like that."

I said, "Okay."

On the morning of the thirtieth day, we were on the hills just outside of Dubrovnik. We could see Dubrovnik. We could see the Adriatic, the offshore islands. One of them was Vis. It was occupied by the British and was Tito's headquarters. We were so excited that we could hardly sleep. That evening we finished our wake-up ritual of fig-fig, bread, slivovitz, and rolled up our blanket. And carefully and quietly walked to a village just north of the town. Maruska had a contact there. He was a fisherman. He had a boat. Maruska found him and explained who we were—Americanski. We got in this sailboat about twenty feet long and we set out. About five hours later, we were at Vis. We were safe. We had made it. We had made it!

We told them who we were. Everybody congratulated us. The British questioned, debriefed me, and then the partisans questioned Maruska. That evening, we were given clean clothes, British Army uniforms, and we went to the officers' mess.

And that night Tito gave Maruska a medal called the Order of the Red Star. Maruska wept. The next morning, we said our goodbyes, they put us in a British torpedo boat, we crossed the Adriatic, they took us to the American hospital in Foggia. They checked us out and that evening, Maruska and I spent our first evening apart. She was in the women's venereal disease ward and I was in the men's.

The guy next to me in the VD ward was a Greek Air Force fighter pilot, Stephan Papadakis. He had syphilis. We talked a while. He was from Athens.

Maruska came to visit me. I introduced her to Stephan. She eyed him pretty thoroughly.

I thought about marriage vows. I didn't know what they were, but they went something like "Love, cherish and obey," and "In sickness and in health," and "Till death do you part."

I thought about Maruska and me. That's what happened to us. We hadn't been separated for a moment for thirty-one days. We had walked, talked, eaten, slept, together for thirty-one days. We had fucked, gotten drunk, hid in fear, laughed, cried, shit, pissed, vomited, fought, argued, everything. I had wiped her ass for heaven's sake. Could I ever be any closer to anybody?

I walked over to the women's venereal disease ward. I said, "Maruska, let's get married."

Tears welled up in her eyes. She said, "Frat— you know—I didn't want to die a virgin. And I no longer virgin. I love you, Frat. End of my virginity. But marriage? Frat, you got to be practical. I older than you, by more than twelve, fifteen years. You got no business, no profession. You still a boy. I need a man who make a living. I want babies, big family. But, Frat, we had a hell big adventure, didn't we? I never forget?"

I was sent to rest camp at Capri. A month later, I was back flying with my squadron. I had to finish up those last four missions. One day, I got a letter from Maruska. She had married Stephan Papadakis, the Greek fighter pilot. "Wish me luck," she wrote. I did.

Every year, for about the last fifty years at Greek Easter—that's about three weeks later than our Easter—I get a postcard from Athens, always with the same message—"Helluva big adventure, never forget." I save them.

Three weeks ago, I went to Midland, Texas. Went there for a reunion with three other guys, Johnny Jensen, our old bombardier; Tommy Bishop, the nose gunner; and Carl Miller, co-pilot.

We went there because that's where one of the last B–24s in existence is. We wanted to see it. We went out and posed in front of it. Had our pictures taken, messed around for a while. Then we went back to the motel, went to Johnny's room. He had a bottle of bourbon, we started drinking and telling stories.

After a while, Johnny said, "I'm tired, I'm going to take a Shaunessey," and he lay down on his bed and shut his eyes.

I had forgotten about Shaunessey. Shaunessey was a sweet boy who figured it out. He was a bombardier and he couldn't stand it, but he wanted to do the right thing. He didn't know how to handle the whole fucken mess. He'd get dressed, go to briefing, come down with all the guys, stand by the plane, but he wouldn't get in. He wouldn't get in the plane. We then had to put on his flying clothes, by that time he weighed like maybe a hundred and eighty pounds and there was only this hole by the bomb bay to get into and we had to take him and shove him in there. It was difficult, it would take like three or four guys, a couple of guys pulling and a couple of guys shoving. We got him in the plane. Then Shaunessey would take all the flak suits that we had and we'd make a little bed out of it. He would get down and curl up in like a fetal position and he'd start saying his prayers. Shaunessey was nuts, everybody knew. The Colonel knew about it, Major Thompson knew about it, Captain Connor, the flight surgeon, knew about it. Everybody knew about it, but the thing was that they couldn't turn him in because if they did, they'd have to court-martial him or kill him or something. So Shaunessey had to fly with us. He had to die with us or live with us.

When you get to about twelve thousand feet, you have to put on a face mask because there's no more oxygen. When I flew with Shaunessey, I'd put it on and he would be there counting those beads and saying, "Hail Mary, full of grace, the Lord is with thee. Blessed art thou among women and blessed is the fruit of thy womb, Jesus. Holy Mary, mother of God, pray for us sinners now and at the hour of our death. Amen." He was saying the rosary. Then he would look at me, and say, "Rockets, is today our last day? Is today our last day? Is today our last day?"

What he was talking about was that ten planes would go out and nine would come back. Twenty planes would go out and eighteen would come back. Thirty planes would go out and twenty-seven would come back. There was a ten percent loss. Air Force said that was manageable, that's okay. But at the end of ten flights, if you were still alive you were scared. The odds were against you. You were playing fucken Russian roulette. I couldn't handle it, nobody could handle it. There was Shaunessey, he figured it out, if you kind of go into a catatonic state that's the only way you could deal with it. Shaunessey was like me, he was like everybody, "Is this my last day? Is this my last day? Is this my last day?"

Fifty years have passed. I wake up in the morning, I look at my wife, I kiss her, I thank her quietly to myself. I think, "Is this my last day?"

Look, I'll take my teachers anywhere I can find them. I want to learn the trick. I want to learn the trick of how to do life. Learning how to do death. That's easy.

THE COLONEL

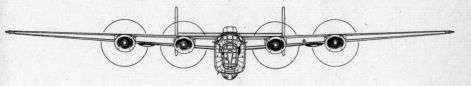

As I remember it, it was a clear hot day in October, October 25, I think. Most of the planes had already landed when Brad's plane circled the landing strip.

Three red flares were shot out, meaning the radio wasn't working, there were wounded on board and have an ambulance ready.

The plane landed; Whitey Wilson, the co-pilot, was at the controls. The plane was full of holes. Three wounded guys were taken off first and were put in the ambulance.

Major Brad Belchore, the pilot, had already been dead for four hours. He had caught a piece of flak over the target, the BMW plant at Munich, Germany.

Everyone on his crew was just white green in shock. The ground crew were crying.

Guys were standing around mumbling. The Colonel came down, looked at the plane, and then looked at Brad's body just lying there. And then he threw up. They said he went back to headquarters and locked himself in his room and got drunk.

Now, I don't mean to say anything bad about the Colonel, he was a hell of a pilot and very intelligent in his own way. But I hated him. I thought he was arrogant, a bully, a bigot, and a coward. I was against everything he stood for. So I was taken aback when his orderly showed up at my tent and said the Colonel wanted to see me.

I went up to headquarters and knocked on his door, he opened it. He looked like hell. He just growled, "Rockets, come down to Bari with me and help me bury Brad. He said you were his oldest friend in the Army. Show me how you say those memorial prayers." I started to protest; he just cut me off and said, "Lieutenant, that's an order."

So I went with him. It was me driving, with the Colonel in front, in a six-by-six truck checked out of the motor pool with Brad in a wooden coffin in back. Bari was where the Army Air Corps cemetery was. Two Italian laborers had dug this hole and were standing around with shovels. They took the coffin and put it on the ground.

The Colonel looked at me to say something. I had this problem. I didn't know any prayers. My parents were atheists and I had never learned any prayers.

I started to talk these words to a song I knew.

So long Brad, it was so good to know you.
So long, it was so good to know you.
So long, it was so good to know you.
But you gotta be leaving this rotten old world.
And you gotta be moving along.
Amen.

And the Colonel just broke down in tears and threw himself on the coffin and cried and said, "Brad, my darling, my darling, my darling," just over and over again.

On the drive back to the base, we didn't talk until the Colonel had me stop at a bar at Foggia. He ordered a bottle of grappa and started tossing down shooters.

After about twenty minutes, he was getting pretty drunk and he got on this kinda talking-crying jag.

He told me how his dad was a mean drunk, a poor sharecropper near this little Mississippi town.

He told me how he asked this gal that he had a crush on to go to the junior prom with him and she said "no," and called him "poor white trash," and he vowed to get even and he managed to get a football scholarship to VMI, Virginia Military Institute, and he decided to make the Army his career and how the war came along just at the right time for him, and when he was still just a lieutenant, he got a date with the general's daughter and he proposed and they got married and the general saw to it that he got regular promotions, and his wife was a wonderful woman and they had two wonderful kids now, a fine family, a perfect family, but perfect as they were, he couldn't stand them. They just got under his skin all the time. They just had him confined, trapped, and he loved being overseas where a man could be free, free.

He grabbed my hand and squeezed it. "You know what I mean, Rockets. 'Free.'"

Then he said, "You don't know how hard it is to find someone who understands you. Someone who is sensitive. Sensitive to your feelings, to your needs. Someone who can make you feel special. You just look and look and look." Then he put his head down on the table between his arms and just moaned, "Ahhhhhhhhhhh...Oh Brad, how I'm going to miss you. I loved you, Brad." And he started crying again. Hard, out of control, like his heart was broken.

I got the Colonel back in the truck and he was quiet again. By the time we got back to headquarters, he had sobered up enough to say, "Lieutenant, you keep your mouth shut about my little, uh, emotional outburst back there. That's an order, you understand? Brad always said you were one guy that could be trusted. Well, keep your mouth shut, understand? That's an order. That's an order."

About two weeks later, an orderly came down from headquarters and handed me a sealed envelope from the Colonel. I opened it.

It was an invitation to go to his place that night at eight o'clock and sit in on a poker game he was having. I knew I had a reputation for being a pretty good poker player, so I thought that was a nice gesture on the part of the Colonel.

I got there and there were four other men and the Colonel, and the six of us sat down at the poker table, then these uneasy discomforting thoughts started coming to me. Like, "Hey, what's wrong with this picture." It was *ME* that was wrong. The other five guys, well, one was a British brigadier general, another a British major, and others were American colonels. I was a lieutenant, what the *hell* was I doing there?

And then everyone was very nice to me, solicitous.

And then they started drinking pretty heavy and telling these dumb, unfunny jokes. And then they'd giggle and laugh too hard, too much. It was phony and unnatural, too tense.

Then the Colonel kept touching me. He put his hand on my shoulder and kept squeezing it. And kept filling my glass with Scotch. It became obvious to me that he was trying to get me drunk.

And time dragged on. The guys' faces were starting to get flushed from the booze. And the damn Colonel kept trying to, you know, sort of cop a feel. Then I realized, "Shit, I am the Colonel's date."

I was ahead on my winnings and knew it wouldn't be smart to leave a winner. So I did some nutty betting and, of course, just then I started getting these sensational, once-in-a-lifetime hands. Like, I drew four natural jacks that I had to fold with. God, what a pain. As soon as I was even, I announced it was getting late and I had to fly tomorrow and started saying, "Good night."

The Colonel's face looked troubled. "You're flying tomorrow?" he questioned.

"Yes sir, with Major Thompson, we're flying lead."

The Colonel let out a growl, "Well, Lieutenant, you can stay here if you want."

"Thank you, sir, but all my gear is down at my tent, and wake-up is at four-thirty and I'd better get moving, but thank you, sir. Good night."

And I stepped outside and closed the door and ran to my jeep and got outta there and breathed a sigh of relief.

It was about midnight. It took about twenty minutes to drive back to my tent. I had the blackout lights on. It was dark. I was sure that I had seen some other lights flicker on the road behind me. "Naw, couldn't be." I was just getting paranoid. I got in my tent, none of the other guys were there. I had just brushed my teeth, taken off my boots, you know, getting ready to hit the sack, and the Colonel walked in.

He came right up to me, put his arms around me, and bent his head down to try to kiss me. I pushed him back as hard as I could. He came at me again, and grabbed me and said, "Oh, you're playing hard to get." The Colonel was an ex–football player, he outweighed me by a hundred pounds and was a head taller than me; I put up a struggle and a wrestling match developed. He blurted out, "Don't fight it, you know you want it, you little prick-teaser. Oh, so you want to play rough." He was starting to feel me up, putting his hands in the damnedest places. I thought, "This jerk thinks he's seducing me, that I'm going to break out into a fit of sexual frenzy."

We were on the floor of the tent. He was on top of me, I just couldn't move him. The Colonel put his face next to mine and then started slobbering all over my face. And then he stuck his tongue in my ear.

Then he said, "Just this once. Let's do it, just this once. I'll make you a major, just like I did Brad. You know what that'll mean."

Making up my mind has always been my weak suit.

There's a little bit of "puta" in everyone. Doesn't everybody have a price? At what point do we sell out and become whores?

Me, a major. Not bad! It happened to old Bradley Belchore, why not me? I had my future to think about. Politics. Being a bemedaled Air Corps major would be great in Arizona, where I come from. It's a conservative state and they go nuts over this hero stuff.

And see, this bombardier friend of mine, Charlie Smith, had gone to the Wharton School of Business in Pennsylvania and had an MBA and he said, "The secret to life was the risk/reward ratio. Is what you're gonna get worth the price you're gonna pay? Was the risk worth the reward?"

Then I thought about the mechanics of the whole thing. Would I be the fuckee or the fucker?

Then who the hell would know anyway, except the Colonel. Then I realized *I* would know and this was a life decision. That last thought just filled me with a tremendous sense of, well, repugnance, or guilt or repentance.

And I yelled out, "Colonel, look, you gotta understand. I'm not Brad, I'm just not queer, not a homosexual like you guys."

The Colonel's jaw just dropped open. He got up, his eyes narrowed into a squint, and he grabbed me by my throat and he started to shake and choke me.

He said, "You little turd, don't you ever call me a queer. You little shit. You're talking to me, Colonel Selby B. Shoots, and you're calling *ME* a homosexual. If I ever hear you say that again, I'll kill you, you understand?"

I gulped out a "Yes sir," and he stormed out of my tent.

"Phew!" The rest of that night I thought I'm in a hell of a mess and how am I going to get out of it?

The next morning, I flew with Major Thompson.

And the next day, I was supposed to have off.

When the weather was good, we'd fly one day on, then one day off.

But instead, the next day I was assigned. I flew lead again to Munich.

Then the next day to Blechammer—the oil refinery in Germany. Then the next day Ploesti, the oil refinery in Rumania. All tough dangerous targets.

Now, these flights last eight to ten hours a day in the air, with briefing and debriefing. It was a twelve-hour ordeal and I was getting exhausted, so I went to Major Thompson who made the assignments and asked what the hell was going on. And he shrugged his shoulders and said he didn't know, he was just following orders from headquarters.

I thought, "Uh-oh, maybe somebody at headquarters wants to get rid of me." Had my death warrant been signed?

Lucky for me, it rained the next four days. The weather had turned rotten and it gave me a breather and a chance to think.

I had to talk to somebody, but who could I trust to keep his yap shut? First, I thought of Bill Connor, the flight surgeon. Too risky. Bill drank a lot and when you're drunk, who knows what'll slip out?

Then I thought of Father Pietro. When I had helped Connor deliver that baby in Orto Vecchio, they invited us to the christening a few days later and I met Father Pietro there.

The chapel in Orto Vecchio was little. Tiny, simple. The community was dirt-poor and Father Pietro was old and simply lived on the food his flock would give him.

He didn't speak much English, but damn good Spanish, as good as mine.

So I put on my dress uniform, went to the PX and bought about fifty dollars' worth of candy bars. I loaded up with all the C-rations and other stuff I could find, like perfumed soap, and corn flakes and a canned ham—stuff I knew he would like.

I checked out a jeep from the motor pool and drove over to Father Pietro's little church. It was a one-room affair with rough benches for pews, a small altar, and a room in back where Father Pietro lived.

First, I gave him the presents and then we said our "how are you" and "how's everything," the necessary preliminary salutations. Father Pietro, I guessed, was about seventy-five years old. White hair, shiny eyes, ruddy cheeks, and somewhat stoop-shouldered. A cheerful face.

He poured me a glass of wine, we said, "Salud." And then I got right down to my business.

"Look," I told him in Spanish, "I've got this problem . . ." And I told him everything. He listened attentively, sympathetically, he would shake his head from time to time and go tch, tch, tch.

He poured himself another glass of wine and said, "My son, God works in mysterious ways. You must remember, never to get into a fight with a man who has nothing to lose. And that in love and war, the only bullets that count are the ones that hit you.

"Also remind yourself that it is easier to pull a chain than to push it. And never set the trap in the sight of the rabbit."

What the *hell* was he talking about? No wonder my parents became atheists.

He went on. "My son, God works in mysterious ways."

I thought, "Yeah, yeah, I know, get to the punch line already."

"My son, one door closes, another door opens up. God gives to some and not to others. Just the other day, Emilio Mazza came and spoke to me, looking for advice, not in confessional, so I can talk to you. Emilio lives in Foggia and is a fine artist in the classic tradition. He is a talented portrait painter. And he was telling me about his problem. It is similar to your colonel's. God decided that Emilio was not for marriage to women, and Emilio is very, very lonely. He wants love and tenderness, like everyone else. Foggia is very conventional and life is difficult for him. And he is looking and looking for someone. That's why he came to me. He knows that I know many souls. Emilio is quite handsome, a nice personality, and speaks a bit of English. Can you arrange for a meeting between Emilio and your colonel?"

I said, "Boy, oh boy, can I ever. Just give me Emilio's address." I said goodbye to Father Pietro. I gave him one hundred dollars. I figured he just might've saved my life.

The rest was easy. I went to see Emilio Mazza. He was a hell of a painter. I worked out a fee that I would pay him to paint the Colonel's portrait. And then I wrote a note to the Colonel that said that I was simply doing what Brad had asked me to do, that is to find a fine portrait painter, and it was a gift from Brad to the Colonel and Brad had paid for it and all the Colonel had to do was to sit for the artist. And I gave the note to Emilio to give to the Colonel and that was that.

I never heard another word. About two months later, I had finished my tour of duty and was on the way back to the United States.

Before I left, I checked in with Emilio and he said the portrait was coming along just fine and the Colonel had ordered some more paintings—wall paintings, murals and such, and he was sort of an "artista in casa," an artist in residence.

By 1950, I had gotten out of college, was married, and had a child. I was working as an apprentice for an architectural firm that was paying me a dollar an hour. My wife and I had all the usual family expenses and were in pretty deep debt.

This firm sent me out to the Burbank Airport to pick up someone. I went into the cafe there for a Coca-Cola and there in a booth was the Colonel and a woman and two young boys. He had put on weight, his hair was grayer, he had on dress uniform with all his combat ribbons. He was a full colonel now.

The woman, I guessed his wife, was chewing him out about something and he looked miserable.

I walked by his table, nodded, and gave him a half salute. He glanced at me, then grimly looked straight ahead.

On the drive back to the office, I thought about my feelings. The Colonel and I had been through a lot. Why did I feel so rejected?

Risk/reward ratio. Had I thrown in a winning hand, four natural jacks, for nothing. What did I want anyway?

MY STRONG
FEELINGS
OF FAMILY

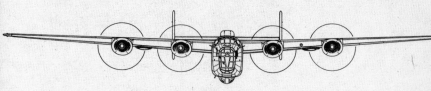

1945

"Uncle Lazar, this is Fred."

"What?"

"Is this Lazar Cohen?"

"Yes, Lazar Cohen."

"This is Fred, your nephew. I'm your sister Rachael's son—Rachael!"

"My sister Rachael?"

"Yes."

"From Arizona. You're calling from Arizona?"

"No, I'm calling from New Jersey."

"New Jersey?"

"Yes, I'm in the Army."

"The Army? So what's wrong?"

"Nothing's wrong, I'm close by so I want to come and visit you."

"Visit me? Again, who are you? What do you want? Speak louder."

I was phoning from the Officers' Club at Camp Kilmer. All this time, I was attracting a lot of raised-eyebrow attention. It was embarrassing. I thought, "Oh, what the hell." I yelled, "I'm Freddie, Rachael's youngest son. You were at my circumcision, remember?"

"Ah, Freddie. I was at your circumcision. I held you. Sure, Freddie, I remember. Where are you?"

"I'm here in New Jersey at Camp Kilmer near Newark."

"Newark. So, why don't you come and visit? We got lots of room."

"Exactly." I answered.

I got off the Greyhound bus at Teaneck, New Jersey. It was bitterly cold. I waited inside the small bus station.

I had just that week returned from Italy. I was twenty-one years old, too many things had happened to me. I felt lost. I have always had a strong feeling of family. I needed some kind of solid touchstone at that time and I needed it badly. That's why I called Uncle Lazar.

In about a half hour, a beat-up truck showed

up, out stepped this husky older man. It was my uncle.

In the cab of the truck sat these three bundles and he introduced me to one of them, his wife, Rebecca. The other two bundles were her little kids. I thought them to be about one and three years old judging from their size.

Lazar explained he lived sixty-five miles away, he raised chickens and he came to town especially early, not only to pick me up but to get a load of grain; wanted to kill two birds with one stone, the back of the truck had a canvas tarp–covered load and the springs of the truck were flat. It was getting darker and colder. Steam formed in a cloud when we talked.

For a second, Lazar and Rebecca stared at me. I was there in my air force officer's uniform, my combat ribbons, my silver wings, my gold hash marks, my silver bars. "He's beautiful," Rebecca blurted. "And I held him at his circumcision," said Uncle Lazar. We had an awkward couple of minutes of talking and then he said, "Get in, get in. We have to drive an hour and a half to get to the house and it might rain yet and the truck don't run too good and if I get a flat tire with this kind of load, I don't know what I'll do because I can't move it, so get in, get in." I started to squeeze in the front cab but it was impossible. The truck was a 1930 Chevy, in the middle was this huge gearshift box with its over-drives. Just barely enough room for Uncle Lazar and Rebecca and the babies. I tried getting in and her sitting on my lap, she was a small woman, but then there was no place for the babies. It just wouldn't work. "It's no problem," Uncle Lazar said. "Rebecca will ride in the back of the truck with the grain and you'll ride up front with me."

He rushed out and untied the tarp and then yelled for Rebecca. We all got out, Rebecca holding the babies. What they had in back of the truck were these gunny sacks of grain, but they were stacked, sort of in a U-shaped log cabin style, leaving an empty spot in the middle about three feet wide and seven feet long, open on one side with the grain sacks stacked about four feet high. Uncle Lazar had a stack of flat gunny sacks laid out in the middle covered with a rough woolen khaki blanket. "Rebecca, jump in with the children."

I resisted, "Look, I'll ride there. I need the rest anyway. I'll just lie down and go to sleep." Now Rebecca protested, but I was adamant. I got in back, Uncle Lazar covered up the load with the tarp, the car wheezed and coughed to a start, and off we started. It was cozy, the dim light seeped in through the cracks in the canvas. But it was bumpy, the springs of the truck bed were flat. I sat for a while. I made a pillow out of a gunny sack, lay down flat on back on the blanket and covered myself with my trench coat. I started to doze off to sleep.

It all happened at once, the truck lurched to one side and a tire hit a bump with an awful bang. The load must have shifted. A gunny sack full of grain slid down my legs, rested between my ankles and my kneecaps. I tried to move the sack but couldn't, it was too heavy. I couldn't get my leverage. I wasn't hurt, but I was pinned down, I couldn't escape. I started yelling, "Uncle Lazar."

At first they didn't hear me. I felt foolish yelling and even sillier that my feet were pinned.

"Uncle Lazar," I kept hollering. Finally the truck lurched to a stop. Uncle Lazar came back. "What do you want."

"A sack has fallen on me and I can't get my feet out."

"Oh! Oh! Rebecca come and help."

Rebecca gave Lazar the babies, hopped in, and struggled with the sack, no movement. Then Lazar handed me the babies, and he tried but couldn't make it. The three of us were quiet for a moment, then Rebecca laughed. "Well, let's count our blessings and let's make the most of it," she said, "If I were back here, that's where I would have put the babies and the sack would have really hurt, maybe even killed them."

"And Freddie isn't hurt, nothing broken," said Uncle Lazar.

"When we get home, the hoist will get this sack off you in a second. And we're only about an hour from home," said Rebecca and we all started to laugh. Lazar said, "Rebecca, get the schnapps." Rebecca returned with a bottle of clear fluid, "Vodka," and handed me the bottle. I took a swallow. It was raw, but warming. Then he took a belt and then he handed the bottle back and she took a big belt.

Then Lazar said, "Rebecca, you ride back here with Freddie. The babies will ride up front with me and they will soon be asleep. Talk to him and keep the vodka, but leave some for me. I started to protest but in hopped Rebecca. Down came the tarp, dark again in this cozy spot and off we went.

As soon as her eyes adjusted to the dimness, Rebecca started her housekeeping building a nest. She straightened out the gunny sacks and made a better pillow with another blanket that was in the back that I hadn't seen. After taking a moment to admire the fleece-lined trench coat, she spread it over me and sort of tucked me in. There wasn't much room. She lay down alongside me, propped her head up with one arm, and stared at me. First quiet and then she said, "You're really beautiful, like an angel."

I had never heard that before. She explained she had only been in this country a few years. She managed to get out of Poland just in time by crossing through Russia and Siberia and slipping into India of all places. "A real adventure," she said. By the time she got to New York, she was penniless, but strong, and after finding a job she looked up the "society" of people of the village she came from. There she met Lazar Cohen, he was from the same town and looking for a wife. He proposed the next week. She accepted and happily moved to New Jersey. He had a nice-sized chicken farm. He made a good living. He bought a wonderful house, in this little town called Arlington so they wouldn't have to live with the chickens. "Lazar is a good man," she said. "Steady, hardworking, dependable." She said she loved him not at first, but little by little, the best way.

And they had lots to eat, and the babies were healthy. As she spoke, her eyes sparkled and her teeth shined as she smiled. We sipped on the vodka bottle. "Let's drink what we want. I have another bottle in front I always carry an extra for emergency. I learned that in the old country."

"How old are you?" I asked.

"Twenty-eight. I'm glad you came. I need family," she said, "All I can get." All her family, she thought, was probably dead by now, she hadn't heard from anyone; she kept praying for a miracle. Lazar told her that he had read that everyone in their village had been killed by the Germans, rounded up, machine-gunned, and buried in a common grave. "It's terrible," she said quietly. "I'm so lucky to be alive." She took a quiet sip of vodka and stared at me, "Freddie, you are so beautiful. I'm going to ask you for something silly."

I said, "Okay, ask." I wasn't drunk, just cozy and mellow and I enjoyed listening to her.

"I want to give you a kiss," she said, "a real one." Then she giggled.

"Rebecca, wonderful, the pleasure is mine."

With that small permission, she leaned over, ran her hand through my hair, cradled my head in her hands, God, I could feel their strength. She kissed my eyelids, my cheeks, gently brushed her lips against mine, brushed her tongue around my lips, separated my lips, and danced her tongue next to mine. After a moment, we both gasped for air. She stared into my eyes as if to give herself justification and said, "I'm cold, I must get under your coat."

And cold she was, she lay there next to me shivering. I put my arms around her and drew her close to me.

We were both slow and passed time as if exploring an unknown, perhaps forbidden land. Not knowing how far to go and not having signposts telling us the direction. It was cold, we were both dressed to the gills. I had on my officer's coat, buttoned and buckled, and the tie and shirt, and undershirt. Her hand dug around until it found its way through the clothing and she stroked my chest. After a bit, she dug into my pants. All this time, my legs were still pinned. She got on top of me and I don't know how she did it, but there she was and here we were in the back of this bouncy truck in the dark covered by this tarp in our secret little house and she had slipped herself on top of me, she was warm and wet, and I was in her.

Afterward, she straightened herself out, buttoned me all up. "Look, beauty," she said, "we must be practical and we're almost home."

In about fifteen minutes, we did get to the house, Lazar got out and Rebecca took the babies into the house. Lazar lowered a hoist, cranked the bag of grain off my legs. I rubbed for circulation. In the next five minutes, I had been shown the extra bedroom, it was late, I fell asleep.

About five o'clock in the morning, I heard these

noises and the truck start and drive off. Lazar had gone to work. Rebecca came to me. She brought me a cup of coffee. She got in bed with me. We made love and after we had breakfast. I went for a walk in town. Uncle Lazar came back about four o'clock. We had a big dinner. Lazar got a bit drunk and told stories about my family that I had never heard before.

The next morning we slept in late, till six o'clock. After breakfast we piled into the truck again and drove back to Teaneck. We got to the Greyhound station, I checked and had about a two-hour wait. Lazar wanted to go to the hardware store to buy some stuff; Rebecca and the babies and I waited at the bus station. Lazar said he'd be back in an hour.

Rebecca looked at me and started to talk. "Freddie, you were brought to me by an angel, like a present, a piece of candy. But God gave me Lazar. That's all.

"I'm strong, like iron, like steel. When I first saw you, I started to shake. I couldn't help myself. I knew very deep down what I was going to do. HAH! You had no chance. Did you think I couldn't move those sacks? I could move them easy, easy, but it was a sign from God. The angels brought you to me. The angels are God's messengers, you know. But God sent you to me and not only that, he even trapped you for me. He trapped you like a little bird. Tied your legs, you couldn't escape. God wanted to give me a little present because I worked hard. To turn down God's generosity would be blasphemy. What could be clearer. Look, beautiful, you just don't understand my kind of love. How could you? You haven't been through what I've been through.

"I have had my head pushed into the asshole of death; stuck there like a cork in a bottle; surrounded by pain and hate and black, black shit. A miracle. I didn't give in. I was able to pull out.

"Now I'm with life. In the old country, I had to take my life and smear it with fear, and then cover the fear with the mud of anger. We did all this, just to survive in that unlucky society; here, it is different and slowly, slowly, here in Arlington, New Jersey, I can wash off some of the fear and anger and touch my life again.

"I don't trust rules! I don't trust rules! They push you down and step on your life. They use cleverness and violence to impose their will.

"I am learning to love again. You know, hate ends only with love. For a long time, in my head was only confusion. Now all is crystal clear, all of the time.

"You know, beautiful Freddie, in the old country we studied the Talmud a little bit. We talked about good and evil and the continuous fighting between the two. For me, the fight is over; for me, I know, there is only good. With good I know I can gamble on the regularity of nature.

"In the old country, when we were running from the Germans, if we had the chance and the strength, we would screw like rabbits, like the desperate rabbits that we were. Rules, morals, shame all thrown to the wind. We were on the edge of death, we wanted a last taste of life. So we would screw and we were right.

"And so when the angels sent me you, like a piece of sweet chocolate, I think it was because I am a good mother and a good wife and I work hard. And you are delicious. I won't forget. But it's not for you that I spread my thighs, it is for life. I know I don't have long to live.

"In my family it is our tradition. We have a bad seed. A disease. We all die at about forty years. I don't know what it is, and how I will hate to leave this life, and Lazar! How I love him. God sent him to me. And I will always work hard and take care of him and I will never betray him. You know Lazar was in the Bible. Lazar, the trusted faithful servant, that's him. He will never let you down."

Uncle Lazar came back from the hardware store. The bus arrived.

I went to shake their hands but they threw their arms around me and kissed me goodbye.

The next day at Camp Kilmer, reveille sounded, the flag waved in the breeze, the sun peeked through the clouds long enough to melt the frost on the ground, soldiers walked around and tried to look busy and preoccupied. Life goes on.

After breakfast, I was drinking my coffee. I knew what Rebecca was talking about.

I was in the same boat. I had been next to death for so long, I was exhausted. Your values change. Like the death camp victims, you want one last chance at what you perceive life is.

ABOUT ACCURATE RECALL

I REMEMBER FLYING FROM DAKAR in the Senegal across the Sahara Desert through the Zagora Pass into Marrakech, Morocco. We were low on fuel. We landed at this dusty town, Timbuktu, mud huts, everyone speaking French. American Air Force fuel depot. Thousands of barrels of fifty-gallon, one hundred octane aviation fuel. We had cold beers. Refueled, took off, flew through the Zagora Pass, through the Atlas Mountains and into Marrakech. I remember all this with pristine clarity.

It never happened. I checked my old navigator logs. We didn't land to refuel. We flew right through the Zagora Pass. And we wouldn't have refueled at Timbuktu anyway. Too far away from the course of our flight. So, where did that memory of that dusty French African town come from?

My memory, it's accurate and false at the same time. It's complex and simple. It changes constantly, often just to fit the circumstance. And yet, all this time I know I'm telling the truth because I'm relying on my memory.

I think one of the most generous things a person can do is share themselves wholeheartedly with another person. How do you do that? You take off your mask and the mask that's under that mask. You reveal yourself. You reveal yourself in the stories you choose to tell. We become redundant. We tell the same story over and over again.

The memory of our memories; the story of our stories. Complex and simple. Sometimes interesting, sometimes boring, sometimes true, sometimes not true. Always revealing.

Listen to

OLD MAN IN A BASEBALL CAP

A Memoir of World War II

as performed by the author

FRED ROCHLIN

"Rochlin is a superb storyteller, and one is easily
enthralled by his performance...no one can deny
that this is a first-rate storyteller in action."
—*AUDIOFILE*

**Winner—Listen Up Audio Award for
Best of the Best 2000**
—*PUBLISHERS WEEKLY*

ISBN 0-694-52241-4 • $18.00 ($27.50 Can.)

3 1/2 hours; • 2 cassettes

Unabridged

Available at your local bookstore, or call 1-800-331-3761 to order

 HarperAudio
An Imprint of HarperCollins*Publishers*
www.harpercollins.com